CAMBRIDGE LIBRARY COLLECTION

Books of enduring scholarly value

Technology

The focus of this series is engineering, broadly construed. It covers technological innovation from a range of periods and cultures, but centres on the technological achievements of the industrial era in the West, particularly in the nineteenth century, as understood by their contemporaries. Infrastructure is one major focus, covering the building of railways and canals, bridges and tunnels, land drainage, the laying of submarine cables, and the construction of docks and lighthouses. Other key topics include developments in industrial and manufacturing fields such as mining technology, the production of iron and steel, the use of steam power, and chemical processes such as photography and textile dyes.

Richard Trevithick

To mark the centenary of the death of Richard Trevithick (1771–1833), H.W Dickinson and Arthur Titley published this fascinating book on the Cornish engineer and his work. They produced an account which appeals to the scientist, the historian and the general reader without over-simplifying the technical details. Best remembered today for his early railway locomotive, Trevithick worked on a wide range of projects, including mines, mills, dredging machinery, a tunnel under the Thames, military engineering, and prospecting in South America. The book and other centenary activities helped to restore Trevithick's rather neglected reputation as a pioneering engineer of the Industrial Revolution, although his difficult personality and financial failures caused him to be overshadowed by his contemporaries such as Robert Stephenson and James Watt. The book places Trevithick's achievements in their historical context, and contains many illustrations of his inventions.

Cambridge University Press has long been a pioneer in the reissuing of out-of-print titles from its own backlist, producing digital reprints of books that are still sought after by scholars and students but could not be reprinted economically using traditional technology. The Cambridge Library Collection extends this activity to a wider range of books which are still of importance to researchers and professionals, either for the source material they contain, or as landmarks in the history of their academic discipline.

Drawing from the world-renowned collections in the Cambridge University Library, and guided by the advice of experts in each subject area, Cambridge University Press is using state-of-the-art scanning machines in its own Printing House to capture the content of each book selected for inclusion. The files are processed to give a consistently clear, crisp image, and the books finished to the high quality standard for which the Press is recognised around the world. The latest print-on-demand technology ensures that the books will remain available indefinitely, and that orders for single or multiple copies can quickly be supplied.

The Cambridge Library Collection will bring back to life books of enduring scholarly value (including out-of-copyright works originally issued by other publishers) across a wide range of disciplines in the humanities and social sciences and in science and technology.

Richard Trevithick

The Engineer and the Man

H. W. DICKINSON
ARTHUR TITLEY

CAMBRIDGE
UNIVERSITY PRESS

CAMBRIDGE UNIVERSITY PRESS

Cambridge, New York, Melbourne, Madrid, Cape Town, Singapore,
São Paolo, Delhi, Dubai, Tokyo, Mexico City

Published in the United States of America by Cambridge University Press, New York

www.cambridge.org
Information on this title: www.cambridge.org/9781108016353

© in this compilation Cambridge University Press 2011

This edition first published 1934
This digitally printed version 2011

ISBN 978-1-108-01635-3 Paperback

RICHARD TREVITHICK

LONDON
Cambridge University Press
FETTER LANE

NEW YORK · TORONTO
BOMBAY · CALCUTTA · MADRAS
Macmillan

TOKYO
Maruzen Company Ltd

RICHARD TREVITHICK

THE ENGINEER AND
THE MAN

by

H. W. DICKINSON and
ARTHUR TITLEY

TREVITHICK CENTENARY
COMMEMORATION
MEMORIAL VOLUME

CAMBRIDGE
AT THE UNIVERSITY PRESS
1934

PRINTED IN GREAT BRITAIN

To

RHYS JENKINS,

at whose suggestion this work was undertaken,
in appreciation of the work he has done in promoting
the study of the history of engineering
and technology

CONTENTS

Preface   *page* ix

List of Plates   xiii

List of Figures in the text   xiv

Chronology of the Life and Work of Richard Trevithick,
    Civil Engineer   xvi

Chapter I. Introductory   1

    Invention—The steam engine—Need of pioneers—Coming of Trevi-
    thick—Guiding principle of his life.

Chapter II. Early Life in Cornwall   6

    Cornwall in the eighteenth century—Character of its people—Mining in
    the county—Streaming—Deep mining—"Cost Book" system—Mine
    administration—Advent of the atmospheric engine—The mine engineer
    —Richard Trevithick senior—Boyhood and education of Trevithick—
    Coming of Boulton and Watt to Cornwall—Trevithick's first employ-
    ment—Tincroft engine trial—Association with Edward Bull—Erects
    engines in opposition to Boulton and Watt—Injunction—Boulton and
    Watt leave Cornwall—Meets Davies Giddy—Death of his father—
    Marries Jane Harvey—Extensive employment as engineer—Plunger
    pump—Water-pressure engine—Experiments with high-pressure steam.

Chapter III. The Field Widens   43

    High-pressure steam—Road carriages, Camborne and London—High-
    pressure patent—Experimental engine at Coalbrookdale—Spread of the
    high-pressure engine—Explosion at Greenwich—Samuel Homfray and
    the South Wales locomotive—Invasion of England—Dredgers for the
    Thames—Thames Archway Co.—London locomotive—Sale of patent—
    Nautical labourer—Iron tanks—Illness—Return to Cornwall—Bank-
    ruptcy.

Chapter IV. Invention at Flood-tide   125

    Wheal Prosper, first Cornish engine—Expansive working—Thrashing
    engines—Steam cultivation—Sugar mills for West Indies—Plymouth
    breakwater—Tour with Rastrick in Cornwall and Devon—Plunger-pole
    engines at Wheal Prosper and at Herland—Cornish boiler—Com-
    pounding engines—Rivalry with Arthur Woolf—Recoil engine—Screw
    propeller—Characteristic conduct.

Chapter V. The Great Adventure   *page* 159

Silver mines in Peru—Arrival of Francisco Uvillé in England—Engines
for high altitudes—Takes shares in Cerro de Pasco Mines—Exports
machinery—Sails for Peru—Jealousy and intrigue—Prospecting for
copper in Chile—Revolution and civil war—Destruction of mines—
Meets Simon Bolivar and Lord Cochrane—Devotion of Mrs Trevithick
—Visits Costa Rica—Journey through forests to Caribbean Sea—Meets
Robert Stephenson at Cartagena—Returns to England.

Chapter VI. Last Flashes of his Genius   207

Changes during his absence in South America—Company promoting,
Costa Rican mines—Gun carriage—Visits Holland on drainage
schemes—Petition to Parliament for grant—Duty of steam engines—
Patents for boilers and superheaters—Application to Admiralty for trial
vessel—Domestic heating—Reform Act column—Work at Dartford—
Illness, death and burial—Character.

Trevithick Portraiture   261

Memorials to Richard Trevithick   264

Letters Patent for inventions   269

Appendix I. Text of Trevithick and Vivian's patent
specification, 1802

Appendix II. List of Trevithick's patents for inventions

Bibliography   281

Pedigree of Trevithick family   *between pages* 284 *and* 285

Index   285

PREFACE

THE life story of Richard Trevithick, one of the greatest
engineers whom the world has known, exerts an attraction
on all those who have made a nodding acquaintance with
the facts of his romantic career, and the more the story is studied
the greater does the attraction become. This at least is the case
with the present authors, for it is now as much as half a century
since the one, and a quarter of a century since the other, was
first attracted to this subject.

Most of the material for the present study has been derived
from the biography by his son Francis published in two large
volumes in 1872. Those who have delved into them must have
felt that, while we are under a great debt to their writer for the
mass of material he has recorded, yet the lack of critical examination of the documents, the many repetitions, and the want of
correlation make the work a most difficult one to master. To
some extent this is due to the fact that the writer subordinated
the chronological sequence of events to the treatment of his
father's inventions one by one.

In the present volume the authors have sought to present a
consecutive story of Trevithick's life, showing under what in-
fluences of events and men his inventions were conceived; upon
what soil, fertile or barren, the products of his prolific brain
fell; what was the immediate, and what the abiding outcome
of his inventions. At first, no more ambitious aim was present to
the authors' minds than to re-arrange and condense the matter
contained in the biography mentioned above, so as to get a
clearer view of the facts for their own satisfaction. They recognized that the writer, while honestly presenting his matter to the
best of his knowledge and belief, smarted under a sense of the
neglect which his father's memory had suffered and put forward
some exaggerated claims, to which his relationship to his subject
contributed. Apart from all this, in the sixty years which have
elapsed since his book was published new matter has been

brought to light, so that altogether we can now get a fairer perspective of events than was at that time possible.

It is of vital importance to bear in mind that adequate documentation for a life of Trevithick exists for about one-eighth of his career only; the rest is largely blank, and particularly is this true of the South American period. It is regrettable that some material in existence in Francis Trevithick's day has since been lost or destroyed. Since all the facts will never be fully known, the authors have endeavoured to reconstruct a balanced presentation of a fascinating personality and of his career, by critical examination of surviving material, some now adduced for the first time; by the evaluation of the motives of action of the men concerned in the story; by weighing the evidence of contemporaries; and by consideration of the state of the arts and of the economic conditions of the time. The result they now submit to the indulgence of the reader; and that they are at all enabled to do so arises from circumstances now to be detailed.

The approach of the Centenary of the death of Trevithick led naturally to the suggestion of publication and, in anticipation of this event, much of this work was put down on paper. The Trevithick Centenary Commemoration Committee, which was set up in October, 1932, to honour the memory of Trevithick by marking the occasion in a suitable manner, decided as one of its aims to publish a memorial volume. The authors offered their work to the Committee and it was accepted. The publication, however, presented difficulties, since the response to the Committee's appeal for funds was insufficient to cover all that it had been proposed to do, until the Directors of Messrs Babcock and Wilcox Limited generously undertook, as their special contribution to the fund, to defray the cost of publication. The authors offer their grateful thanks to the Committee and to the Company for their hearty support, without which it is only too probable that the work would have remained in manuscript form.

The authors acknowledge thankfully help from quarters too numerous to be mentioned individually here, but they believe that they have not omitted to give due acknowledgment in the text. They must, however, record their thanks to those gentlemen who

have given access to the main sources of information: Captain
R. E. Trevithick, a great-grandson of the inventor; the Secretary
of the Royal Cornwall Polytechnic Society; the Secretary of the
Royal Institution of Cornwall; the Librarian of Falmouth; the
Librarian of H.M. Patent Office; the Corporation of the Trinity
House; the Officers of the Science Museum, South Kensington,
and the Council of the Newcomen Society for the Study of the
History of Engineering and Technology.

Lastly, the authors are most grateful to the Syndics of the
Cambridge University Press for having undertaken the publication of the book.

December 1933.

LIST OF PLATES

I. Birthplace of Richard Trevithick, as it appeared in 1871   *facing page* 13

II. Richard Trevithick as a boy   15

III. Holograph report by Trevithick on Hornblower's compound engine, 1792   19

IV. Original drawing by Trevithick of Bull's engine, *c.* 1792   21

V. Original drawing by Trevithick of his water-pressure engine at Trenethick Wood, 1799   41

VI. Trevithick's experimental model road locomotive, 1798   45

VII. Drawing enrolled with Trevithick and Vivian's patent for the high-pressure engine, 1802   53

VIII. Holograph letter from Trevithick to Giddy, 1804   63

IX. Early drawing of Trevithick's tram locomotive, 1803   69

X. Original drawing of Trevithick's Newcastle locomotive, 1805   71

XI. Trevithick's high-pressure engine and boiler by Hazledine & Co., *c.* 1805   76

XII. Plan and section of the Thames drift-way, 1809   105

XIII. Trevithick's home at Penponds, as it appeared in 1870   123

XIV. Trevithick's high-pressure engine and boiler, used for thrashing, 1812   131

XV. Oil portrait of Richard Trevithick, aged 45   159

XVI. Mule path from Lima to Cerro de Pasco through the Andes, present day   173

XVII. Statue of Richard Trevithick at Camborne, unveiled 1932   262

XVIII. Memorial tablet to Richard Trevithick in Dartford Parish Church, 1902   267

LIST OF FIGURES IN THE TEXT

1. Register of Baptisms, 1771, Illogan Parish Church *page* 13

2. Trevithick's water-pressure engine at Wheal Druid, 1799   40

3. Trevithick's water-pressure engine at Alport Mines, Derbyshire, 1803   41

4. Trevithick's high-pressure whim engine at Cook's Kitchen Mine, 1800   45

5. Trevithick's Camborne road carriage, 1802   49

6. Expenses of the patent of 1802 and of the London road carriage   58

7. Trevithick's high-pressure engine and boiler, 1803   62

8. Trevithick's engine for driving puddle rolls at Tredegar, 1804 to 1856   71

9. Trevithick's boiler for the rolling mill at Tredegar, 1806   73

10. Trevithick's dredger on the Thames, *c.* 1806   86

11. Trevithick's high-pressure engine used with the dredger on the Thames, *c.* 1806   88

12. Site of proposed Thames Archway, 1805–1809   91

13. Trevithick's plan for a tunnel of brick, in sections, under the Thames, 1808   103

14. Admission card to Trevithick's railway, 1808   108

15. Site of Trevithick's London railway, 1808   112

16. Trevithick's steamboat with paddle wheel amidships, 1806   116

17. Trevithick's high-pressure engine for discharging cargo, 1804   117

18. Trevithick's boiler with direct contact of fuel gases and water, 1811   126

19. Sketch of arrangement of Wheal Prosper engine, 1811   129

20. Trevithick's thrashing engine at Trewithen, 1812   130

21. Trevithick's steam cultivator, 1813 *page* 133

22. Plug and Feathers for breaking up stone, 1813 138

23. Rastrick's design of a Trevithick engine for factory purposes, *c.* 1813 141

24. Rastrick's design of a diagonal Trevithick engine for a steamboat, *c.* 1813 142

25. Trevithick's plunger-pole engine, 1815 146

26. Trevithick's recoil engine and boiler, 1815 150

27. Trevithick's water-tube boiler, 1815 152

28. Trevithick's plunger-pole engine with water diaphragm, 1815 154

29. Trevithick's winding engine for South America, 1813 166

30. Ledger account with Uvillé 171

31. Map of route followed by Trevithick across the isthmus of Nicaragua, *c.* 1826 201

32. Trevithick's recoil gun mounting, 1827 216

33. Trevithick's ball and chain pump for Holland, 1828 224

34. Trevithick's closed cycle steam engine and boiler, 1829 237

35. Trevithick's vertical tubular boiler, 1830 244

36. Trevithick's apparatus for heating apartments, 1830 248

37. Column to commemorate the passing of the Reform Bill, 1832, proposed by Trevithick 253

38. Register of Burials, 1833, Dartford Parish Church 256

39. Trevithick and Vivian's high-pressure steam engine, 1802 patent 271

40. Trevithick and Vivian's high-pressure steam engine and sugar mill, 1802 patent 274

41. Trevithick and Vivian's steam road carriage, 1802 patent 277

CHRONOLOGY

of the

LIFE AND WORK OF RICHARD TREVITHICK

Civil Engineer

Age	Year	
	1771	April 13th. Born near Carn Brea in the parish of Illogan, Cornwall.
6	1777	Boulton and Watt introduce Watt's patent engine into Cornwall.
19	1790	Employed as engineer at Eastern Stray Park and other mines.
21	1792	Associates with Edward Bull.
		Reports on the engine of Hornblower at Tincroft Mine.
24	1795	Re-erects an engine at Wheal Treasury.
		Erects an engine at Ding Dong Mine, infringing Watt's patent.
25	1796	Visits Soho Foundry, Birmingham, with Bull, when injunctions are served on both.
		Alters Ding Dong engine to work as an atmospheric one.
		Becomes acquainted with Davies Giddy.
26	1797	Richard Trevithick senior dies.
		November 7th. Marries Jane Harvey.
		Makes models of high-pressure stationary and locomotive engines.
27	1798	Puts to work his first water-pressure pumping engine.
		Extends his application of the plunger pump in the mines.
		Makes small high-pressure engines for winding and work on the mines.
		First engine sent to London in charge of Arthur Woolf.
29	1800	Boulton and Watt leave Cornwall on expiration of Watt's patent.
30	1801	Builds his first steam road carriage at Camborne.
31	1802	March 26th. Patents with Andrew Vivian the high-pressure steam engine for stationary and locomotive use.
		William West becomes a partner in the patent.
32	1803	Builds a second steam road carriage and sends it to London.
		Commences erection of his high-pressure engines in London and elsewhere.
		Meets Samuel Homfray and builds engines for ironworks in South Wales.
		Builds an experimental engine at Coalbrookdale.
		Homfray now interested in the patent.
		Explosion of one of his cast-iron boilers at Greenwich.
33	1804	Many foundries engaged in making his engines.
		Builds a tramway locomotive at Penydaren.
		Andrew Vivian sells his share in the patent to Robert Bill.
		Sent for by the Admiralty to explain his engine and meets Simon Goodrich.
		Invasion of England threatened by Napoleon. Suggestion to use steam-driven fireships against the French flotilla.
34	1805	Builds a railway locomotive engine at Newcastle-on-Tyne.
		Drives a canal barge by means of a steam engine and paddle-wheels.
35	1806	Constructs the first of three steam dredgers and puts it to work on the Thames.
		July 22nd. Puts to work a second dredger on the gun-brig "Blazer".
36	1807	Engaged by the Thames Archway Co. to make a driftway under the Thames.
		Homfray holds a controlling interest in the patent.
37	1808	Sells his remaining share in the patent.
		Constructs a passenger locomotive near Euston Road, London.
		Makes the acquaintance of Robert Dickinson and, July 5th, patents with him machinery for towing ships and discharging cargo.
		October 31st. Patents with Dickinson iron tanks for stowage of cargo, etc. on board vessels.

38   1809   April 29th. Patents with Dickinson floating docks, iron ships, iron masts, etc., bending timber, diagonal framing for ships, iron buoys, rowing trunk and steam cooking.
Raises a sunken ship off Margate.

39   1810   Illness and return to Cornwall.

40   1811   Feb. 11th. Made bankrupt. Sequestered for debt.
Sells the tank patent to Henry Maudslay.
Applies himself to the use of steam expansively and builds the first Cornish engine at Wheal Prosper.
Builds his first plunger-pole steam engine at Wheal Prosper.

41   1812   Applies his high-pressure Cornish boiler to the pumping engines at Dolcoath Mine.
Invents the single-acting high-pressure expansive engine and applies it to agriculture.
Supplies an engine and apparatus for rock-boring for Plymouth Breakwater.
Constructs a screw propeller for marine propulsion.

42   1813   Builds an engine for ploughing.
Adds plunger-pole high-pressure cylinders to existing mine-pumping engines, thus compounding them.
Francisco Uvillé arrives in Cornwall from Peru. Trevithick visits him and engages to build engines for the Cerro de Pasco Mines.

43   1814   September 1st. Nine engines shipped to Peru.

44   1815   Builds a recoil engine to be used with a screw propeller for ship propulsion.
Builds a high-pressure plunger-pole pumping engine at Herland Mine.
June 6th (amended November 20th). Patents the plunger-pole engine, recoil engine, and screw propeller.

45   1816   Proposes a tubular boiler.
William Sims buys a half share in the plunger-pole patent.
Deaths of Henry Vivian and William Bull in Peru.
October 20th. Sails for Peru.

46   1817   February 6th. Arrives at Callao. A faction attempts to supplant him at the mines. Death of Uvillé. Reinstated at the mines.
Civil war in Peru.

50   1821   Visits Chile and starts copper mining.
Salves the cargo of a ship near Callao.
Leaves Lima for Bogota in Colombia.

52   1823   Proceeds to Costa Rica where he meets Gerard. They obtain mining concessions there.

55   1826   Gerard and he make a pioneer journey across the isthmus of Nicaragua.

56   1827   They arrive at Cartagena in Colombia and meet Robert Stephenson.
Oct. 9th. Returns to England.
Designs a recoil gun-carriage.

57   1828   Visits Holland in connection with drainage operations and builds pumping-engine for that country.
Draws up Petition to Parliament for a compassionate grant. Meets with financial assistance and drops the Petition.

60   1831   February 21st. Patents apparatus for heating apartments by warm air; a superheater; jet water-propulsion of vessels; and a boiler and superheater applied to locomotives.

61   1832   Proposes huge national column of cast-iron to commemorate the passing of the Reform Bill.

62   1833   April 22nd. Dies at Dartford, Kent.

CHAPTER I

INTRODUCTORY

Invention—The steam engine—Need of pioneers—Coming of Trevithick—
Guiding principle of his life.

I N the history of invention a long time almost invariably
elapses between the birth of an idea and its realization in
practice. Rarely does an invention spring fully matured from
the brain of an inventor, rather does it require additions and sub-
tractions by one brain after another until at last it becomes
practicable. Then, again, the inventor may be before his time.
The harnessing of wind or water is a good idea, but even were
it brought into use, it would remain abortive until the contemporary development of other branches of technique called for the
use of power units larger than the animal. But invention, though
it may be stimulated by economic demands, depends on the progress of the mechanic arts to such a stage that it becomes possible
to reduce the invention to practice. There are really two aspects
to this that act and react upon each other: one is concerned with
the making of the apparatus, the other with the means for obtaining desired results; one is concerned with the machine and
the other with the tools, the latter giving rise to further invention. It is upon the second of these aspects that depends the
length of time that an invention passes in the embryo stage. Incidentally it is not generally realized how much the constructive
arts have owed to war. What is too costly for peace purposes can
always be afforded for war and later on becomes everyday practice. Thus the production of small arms on the interchangeable
system led up to the possibility of the typewriter. The impracticable of one age becomes the actual of another.

The first aspect of the mechanic arts may be instanced by the
case of the long-cherished dream of the mechanically propelled
vehicle. This had to await the invention of the high-pressure steam
engine. The search for more direct methods of converting energy
into work led to the development of the gas engine and, later,

its congener the petrol engine. The petrol engine in turn enabled the aspirations of many a would-be Icarus to be realized in the aeroplane. In the latter cases the tools of the constructive arts were developed and ready to hand; hence phenomenal progress in a short time.

Taking the instance of the mechanically propelled vehicle and considering it in greater detail, since it will help materially in grasping the main thesis of the present volume, we find that it is necessary to go back two thousand years, at least, for it was then that Heron of Alexandria described a sphere which was caused to revolve by steam issuing in jets against the resistance of the air—the prototype of the reaction turbine. Branca also in the seventeenth century illustrated a wheel rotated by jets of steam impinging upon its vanes, i.e. the impulse turbine. These toys, for they were hardly more, relied for their action on steam at a pressure above that of the atmosphere. They could only remain toys till it became possible to construct boilers that would withstand safely the pressure needed to operate them. It was not that the effect of heating water in a closed vessel was not well understood. For example, Edward Somerset, second Marquis of Worcester—and he was not the first— tells[1] how he took

a piece of a whole Cannon whereof the end was burst and filled it three quarters full of water stopping and scruing up the broken end; as also the Touch-hole; and making a constant fire under it, within 24 hours it burst and made a great crack.

Instead of trying to restrain this giant force, as knowledge spread and it became known that the atmosphere had weight, men's ideas turned into another and easier method of approach. Thomas Savery raised water in an apparatus in which he made use of both steam and atmospheric pressure, the latter obtained by the condensation of steam in a closed space. The steam pressure required was considerable, and the difficulties he met with in constructing his steam vessels or boilers were so great that his success was very limited. Thomas Newcomen relied solely on the pressure of the atmosphere obtained by the condensation

[1] *Century of Inventions*, 1663, § 68.

of steam under a piston in a cylinder and thus brought into being the first practical steam engine. James Watt, although he made enormous strides by condensing the steam in a vessel separate from the cylinder, yet did not change the principle of the engine, which still remained a low-pressure apparatus. Watt, great man of science and brilliant inventor, was content to leave the engine at this stage, hence he never faced the difficulties of constructing boilers to resist pressure. He found the boiler merely a hot-water tank and left it very little better. That it was such, and faulty at that, is shown by the fact that it was common practice to lay a rope yarn between the plates before riveting them together to ensure tightness.

Suppose now a boiler is made capable of withstanding a moderate pressure above the atmosphere, the pressure below the atmosphere can then be dispensed with, and with it goes not only the water supply necessary for condensation but also the complication incidental to clearing away the condensed water, i.e. the air pump; the steam, after it has done its office, is discharged into the air. Now no longer tied to the earth and relieved of much of its weight, the high-pressure engine, for such it has now become, can be mounted on wheels and the long-wished-for locomotive becomes an actuality.

Man in the aggregate is a conservative animal. He clings to rule-of-thumb methods which, after all, are the subconscious accumulated experiences of his forebears. Progress would be slow even when circumstances demand it to be rapid if occasionally there did not arise a man—seer, prophet, genius, pioneer, call him what you will—whose vision, originality, or energy is sufficient to force on development against the inertia of his day.

Such a man was Richard Trevithick, who appeared on the scene towards the close of the eighteenth century, when the period of Watt's unduly prolonged patent was drawing to its close. He had the true mechanical instinct which goes to the core of a problem and sees the best way to attack it; he bubbled over with ideas, he was a bold and original experimenter, while nothing could withstand his tremendous energy. In spite of the limited theoretical knowledge of his day, he grasped fundamental principles and hence visualized the right line of advance. While other men would be revolving an idea in their minds, he, with apparent recklessness, would be making an experiment under the severest possible conditions, those of everyday work. Particularly was this the case with the high-pressure engine, with the development of which his name must ever be especially associated. As we have said, he found the steam boiler capable of withstanding only a few pounds pressure above the atmosphere; he constructed boilers to work at 150 lb. and more, not of one type only but of many, but here he went ahead of the constructive arts of the day. However, one boiler with which his name is associated—the Cornish boiler—with its lineal descendant the Lancashire boiler, which is identical in principle and came into vogue as boiler sizes increased, is still a standard design.

The work done by Trevithick in the application of high-pressure steam to stationary and locomotive engines, as well as his activities in many other directions, will appear as the story unfolds. Had he not spread his energies over so wide a field, he might have achieved more success in a part of it. It is this which has perhaps unduly overshadowed the work that he actually did. Till the very end, worldly success eluded his grasp, but he did not allow this to embitter him. He summed up his reflections on his life's work in a statement, without date, but obviously written a short time before his death, concluding with a noble declaration which might well serve as his epitaph:[1]

I have always consulted him [i.e. Davies Giddy] on every thing new that I have brought before the public, and though his practical knowledge could not assist me, yet his theory has been very correct and essential. Nearly thirty years I have been contented with steady, hard labour, and immense expenses, entirely alone, for the great, and even incalculable benefit of my country, without receiving any reward; but have been branded with folly and madness for attempting what the

[1] *Enys Papers. Life of Richard Trevithick, with an account of his Inventions*, 1872, II, 395. An asterisk intimates that the "eminent scientific character" was understood to be John Isaac Hawkins (1771–1855), civil engineer and patentee. What we know of the character of James Watt junior, coupled with the above date, suggests that it was he rather than his father who made the remark.

world calls impossibilities; and even from the great engineer, the late Mr James Watt, who said to an eminent scientific character still living that I deserved hanging for bringing into use the high pressure engine; this so far has been my reward from the public: but should this be all, I shall be satisfied by the great secret pleasure and laudable pride that I feel in my own breast from having been the instrument of bringing forward and maturing new principles and new arrangements, to construct machines of boundless value to my country: and, however I may be straightened in my pecuniary circumstances, the great honour of being a useful subject can never be taken from me, which to me far exceeds riches.

CHAPTER II

EARLY LIFE IN CORNWALL

Cornwall in the eighteenth century—Character of its people—Mining in the county—Streaming—Deep mining—"Cost Book" system—Mine admini-stration—Advent of the atmospheric engine—The mine engineer—Richard Trevithick senior—Boyhood and education of Trevithick—Coming of Boulton and Watt to Cornwall—Trevithick's first employment—Tincroft engine trial—Association with Edward Bull—Erects engines in opposition to Boulton and Watt—Injunction—Boulton and Watt leave Cornwall—Meets Davies Giddy—Death of his father—Marries Jane Harvey—Extensive employment as engineer—Plunger pump—Water-pressure engine—Experi-ments with high-pressure steam.

CORNWALL, where our story opens at the end of the eighteenth century, was, as it is to-day, occupied by a race of Iberian and of Celtic origin engaged mainly in the pursuits of agriculture, mining, fishing and seafaring. In that part of the country the sea and the land are so closely inter-mingled that interchange of occupation 'twixt one pursuit and another is a feature that was quite as thoroughly marked then, if not more so, than at present. The quickening of the spirit that such diversity seems to engender gives us the alert, hardy, re-sourceful population, albeit conservative, clannish and super-stitious, that we see to-day; the character that Milton, in 1644, attributes to the English nation—"not slow and dull but of a quick ingenious and piercing spirit, acute to invent, suttle and sinewy to discours, not beneath the reach of any point that human capacity can soar to"—might seem to have been written with particular reference to the Cornish people.

Distant as is Cornwall to-day from the metropolis, in the eighteenth century it was so remote that communication by wheeled vehicle hardly existed—the pack-horse and the saddle-horse were the means used.

However, the county abounds in natural harbours; Portreath and Hayle on the west, Penryn, Perran Wharf (on inlets of Falmouth Harbour) and Penzance on the south, were all within easy reach of the mining districts; hence practically all goods coming to or going from Cornwall were dispatched by sea.

Falmouth was the headquarters of the packet service to the North American Colonies and the West Indies, as well as to the Mediterranean. The impression that we gather of Cornwall at this period is that of a busy and thriving county.

It is, however, only of one side of the activity of its inhabitants —that of mining—that we have here to deal, and to understand the position at the period of which we speak, a few words are necessary.

Mining for tin has been carried on in Cornwall for over two thousand years. The form that it took throughout the first fifteen hundred years was that known as "streaming", i.e. working alluvial deposits by subjecting the "dirt" to a stream of water which carries off the lighter particles of useless mineral, leaving behind the heavier tin-stone or sand. It was not until the alluvial deposits showed signs of exhaustion that attention was turned to the source of the tin-stone: by "shoding" or following up the surface stone or deposit, the outcrop of the vein or lode bearing the tin-stone was discovered. It was, then, about the middle of the fifteenth century that underground mining proper began, and the streamer became the miner. The methods of working the lode—by open trenches or "goffens", by adits or levels along or across the lode, by sinking shafts down to the lode—were all developed, but not without outside assistance. Such mining was not carried on long before that great bugbear of the miner—water—was encountered. To get rid of it, resort was had to the then known methods, by baling and by chain pumps actuated by animal power or by water-wheels. To raise the vein-stuff as well as the "attle", "deads" or waste mineral, re-quired the provision of windlasses and horse gins or whims. Hence gradually the miners acquired increased mechanical skill and some differentiation in employment resulted among them.

Deep mining called for much more capital than the "streamer" had at his command, so gradually the method of mining by the "Cost Book" system, so typical of Cornwall, grew up. Under the Stannary Laws an immemorial privilege known as "bounding" was enjoyed by the tinner, and consisted in the right to enter upon land to search and dig for tin; if the land was unenclosed, the tinner was privileged to work without paying royalty, but if enclosed some "toll tin" or "toll dish" of perhaps one-fifteenth of the product was payable to the landowner. Bounding by the eighteenth century had become most complex owing to the number of claims set up, and they were irksome to the industry owing to the difficulty in reaching agreement with the several "bounders" when fresh developments were proposed.

Assuming that these preliminary details had been adjusted, a group of men could adventure together to start a mine. They might be "out-adventurers" who simply put in capital, or "in-adventurers" who were merchants interested in the supply of stores. Incidentally it is characteristic of Cornwall to find a man joining in many businesses and engaging besides as a merchant in a very miscellaneous range of goods. Six or more such adventurers agreed to form a "Cost Book Company". At their first meeting the Cost Book was produced, resolutions stating the purpose of the company were entered, and a committee of management (without pay), a purser, manager and engineer (with stated pay) were appointed. The manager was authorized to carry out certain work, not to exceed a certain cost, before the next meeting. The Cost Book was signed by each adventurer against the number of shares he had taken up. The purser took the Cost Book to the bank, the proprietor of which, if satisfied, agreed to honour drafts or orders drawn by the purser to the stated amount. At the next meeting—and these in Trevithick's day were held every two months—the Cost Book in which the purser had entered all wages, merchants' bills, royalties, etc., was produced. The purser also wrote out a balance sheet, on which the amount due to the bank was shown. A call of the necessary amount per share was made to discharge the debt, or if profits had been made, these were distributed in the same proportion. Fresh resolutions were passed and entered and the Cost Book again signed as before. Adventurers who did not attend the meeting were nevertheless bound by its decision, but anyone could relinquish his shares, subject to his liability, up to the time of the next meeting.

A drawback of the Cost Book system was that liabilities could be incurred, e.g. for materials supplied, without their being paid for. A creditor would then draw the company into the Stannary Court, which would make an order on the adventurers. Should any of them be unable to pay—the liability being "jointly and severally"—the remainder had to pay. On the other hand, an advantage of the system was that capital was not subscribed till wanted and a close hold over expenditure was maintained, so that at any time a mine could close down quickly should it become non-paying.

The manager of the mine was responsible for the selection of the working staff, principal among whom were the captains— miners of great practical experience, to whom the actual working of the mine was entrusted. They made the bargains with the miners for their pitches when working as "tributers", or for getting out ground in shafts and levels when acting as "tut-workers". There were underground and "grass" captains, the latter being responsible for dressing the tin or copper ore after its delivery "to grass", i.e. on the surface. The captains kept the plant in order. On their skill, acumen and integrity depended the success of the mine.

Other important workers were the "binders" or timbermen, the "bal-carpenters" and the smiths. Many more, even to "bal-maidens"—and children too on light jobs—at the surface, were employed, indeed a mine was not only an industrial unit but even a social centre.[1]

The amount of wages earned by the persons employed may be judged from monthly figures taken from the Cost Book of the "Great Work in Breage", 1759–64: Purser 50s., Captains 40s., Binders 42s., Carpenters 27s. to 36s., Smiths 30s. to 35s., Engine-men 10s. to 12s. The miners earned from 16s. to 21s., with frequently "extra stems" or shifts of 1s. to 1s. 6d.

[1] See *Trans. Newcomen Soc.* xi, 26, Titley, "Cornish Mining". Hamilton Jenkin, *The Cornish Miner*, 1927, treats in detail the subject baldly set forth above, and should be consulted.

In addition to selecting the staff, the manager had also the task of selling the tin or copper ore produced, and for this purpose attended the "ticketings" held alternately in Truro and Redruth. Although the individual mine offered its ores for sale once in four weeks or even once in eight weeks only, the mines were so grouped in districts that there was a ticketing practically every week—an obvious convenience for samplers and buyers alike. The ore produced in the mine was carefully weighed off and piled into "doles", i.e. heaps of circular form with a flat top, all of the same weight—perhaps 10 tons—and of the same quality of ore. A sampling day was appointed and notices sent to the smelting companies; they sent agents, who could if they chose take a sample from, and weigh, any two of the doles. The agents had the samples assayed to find out their value. Managers and agents assembled on the ticketing day in an inn at one of the places named, the chair being taken by the manager whose mine had the largest amount of ore for sale. Each agent wrote on a slip of paper or "ticket" the price he was prepared to pay; as each parcel of ore was reached, the chairman read out the various bids and the highest bidder was declared the purchaser. Business was thus concluded with great dispatch, scarcely a word being spoken.

As mining went deeper and the volume of water encountered was such that horse gins or whims and water-wheels were incapable of dealing with it, a new source of power for pumping had to be sought. The need was great, but fortunately the remedy was near at hand. Thomas Newcomen, an ironmonger of Dartmouth, travelling amongst the mines in the course of business to sell tools, reflected on the need of the hour, and in the opening years of the eighteenth century developed his atmospheric pumping engine, known among the miners as the "fire engine" to distinguish it from the older horse engine and water engine. As its name indicates, it relied upon fire to produce steam which, when condensed under a piston, caused the pressure of the atmosphere to do the work of pumping. Tradition has it that the first engine in Cornwall was erected by Newcomen himself, about 1715, at Wheal Vor in the parish of Breage, but the evidence is inconclusive. We do know, however, that from the arrival in Cornwall of Joseph Hornblower from Staffordshire in 1725, to erect an engine near Truro, the powers, the construction and the operation of the atmospheric engine became gradually known. A fillip to the use of engines was given in 1741 by the remission of the duty on coals consumed in mine drainage in the county.

Borlase reports[1] an increase in the number of engines from one in 1741 to at least a dozen in 1758. Pryce,[2] twenty years later, says that sixty had been built and this is probably an understatement of the actual fact. It should be remembered that, at this period, erecting an engine was like building a house. The materials had to be assembled on the spot mainly from a distance and largely by sea; for instance cast-iron work such as the cylinder came from Coalbrookdale or South Wales; wrought iron from Staffordshire and London; timber beams from the Midlands; copper sheets and pipes from Bristol. Smith's work appears to have been done locally from the beginning and Cornish smiths early acquired repute for skill in their craft. It is to be remembered that there was practically no fitting, i.e. smith's work was to finished sizes. Brass foundries existed early in Cornwall, possibly because a low melting temperature only is required.

With the engines grew up a new race of men who designed and erected them. One of the earliest engineers to reside permanently in Cornwall was Jonathan, the son of the aforementioned Joseph Hornblower, who arrived in 1744 when he was twenty-eight years of age. Several of his numerous sons became engineers. The practice was, it seems, after the engine had been erected, for the engineer to be retained at perhaps 30s. to 40s. per month to keep it in order; thus one man might have engines on several mines under his charge. Another practice was to put the engine in charge of the mine manager or even of the purser, with an allowance to him for his trouble.

We have emphasized the important function served by the

---

[1] *Natural History of Cornwall*, 1758, p. 173.
[2] *Mineralogia Cornubiensis*, 1778, p. 7.

mine manager in carrying on the mining industry. Ability, integrity and force of character, especially the latter, were essential to fill the position. One of the best known of such men in Cornwall in the last quarter of the eighteenth century was Richard Trevithick[1] senior. Beyond that he was born in 1735 we know little of his history and nothing of his antecedents. From some account books that have been preserved we learn that in 1776 he occupied the position of manager at Roskear (Wheal Chance) and in the following year he is recorded as manager, in addition, of Dolcoath, Wheal Treasury and Eastern Stray Park Mine at 40s. per month for each. Francis Trevithick states[2] that he was agent for Lord Dedunstanville from 1777 till his death. As such probably he attended the ticketings of ore at Redruth, the record of which appears in another of his account books. Giddy, speaking of the son, says

His Father was the chief Manager in Dolcoath Mine and he bore the reputation of being the best informed and most skilful Captain in all Western Mines; for so broad a line of distinction was then made between the Eastern and Western Mines [the Gwennap and the Camborne Mines] as between those of different Nations.[3]

We are told that Trevithick senior was a follower of John Wesley and a class leader; also that Wesley, who visited Cornwall in 1753 and 1781, was entertained at his house, but there is no entry or mention of this in Wesley's *Journal*. Wesley's influence in Cornwall, as in other places, was very great; by his preaching and example he raised the standard of personal conduct and established the system of local preachers to carry on the work he initiated. An annual Wesley memorial service at Gwennap has been held every year up to the present.

Richard Trevithick senior married in 1760 Anne Teague, one of a well-known family, many of whom were mine managers in the Redruth district. Their union was blessed with four daughters and one son Richard, destined to become the famous

[1] Outside Cornwall the authors have found doubt as to the proper pronunciation of the surname; the accent is on the *second* syllable "vith".
[2] *Life*, I, 30.
[3] *Enys Papers*. Letter, Davies Gilbert to J. S. Enys, 1839, April 29th.

PLATE I. BIRTHPLACE OF RICHARD TREVITHICK,
AS IT APPEARED IN 1871

Courtesy of the Cardiff Public Libraries

engineer to whose life and work this volume is devoted. Richard, born April 13th, and baptized May 12th, 1771, was the youngest child but one and the only boy, so that it is not to be wondered at that under these circumstances his mother spoiled him.

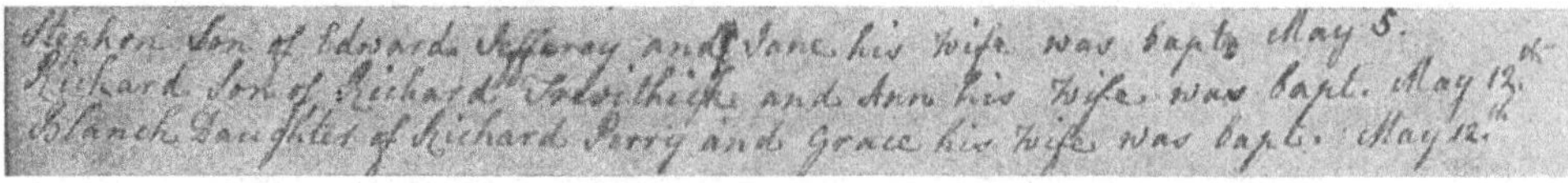
Stephen Son of Edward Jefferay and Jane his Wife was bapt. May 5.
Richard Son of Richard Trevithick and Ann his Wife was bapt. May 12th
Blanch Daughter of Richard Perry and Grace his Wife was bapt. May 12th

Fig. 1. Register of Baptisms, 1771, Illogan Parish Church

At the time of his birth the family was living in the parish of Illogan within easy reach of the mines which his father managed; the children were baptized in the parish church and their names appear in the Register of Baptisms (see Fig. 1).

Francis Trevithick says[1] that the "manager's residence" was "delightfully situated at the foot of the north-west slope of Carn Brea" and that it is "now [i.e. 1871] a double cottage around which clouds of mineral sand from the surrounding mine-works float in the wind". In the illustration which he gives—a woodcut by W. J. Welch—he is not so explicit because he entitles it "Residence of Trevithick sen. in 1760, as it appeared in 1871". We possess the photograph from which the artist obviously worked, together with another photograph (see Plate I) of the same spot; the latter has the inscription "House in which Richard Trevithick was born in 1771. Parish of Illogan". This photograph was formerly in the possession of Francis Trevithick and is now in the Cardiff Reference Library.[2] With the aid of these photographs and his local knowledge of the topography, Mr W. A. Michell has identified the spot as being on the south side of Pool village close to South Wheal Crofty Counthouse, near where there are enormous heaps of waste sands from the stamps. On this spot is a house, known as Penhellick, built in 1881 by Capt. Zacharias

[1] *Life*, I, 52.

[2] For facilities to reproduce it we are indebted to the Librarian, Mr Harry Farr.

Williams and now occupied by his son, Capt. Dick Williams, whose description of the cottages previously existent tallies with that given above. On the other hand, there is a local tradition that the birthplace was a house, the site of which is now occupied by Tregajorran Chapel. These two sites are less than half a mile distant from one another. It is a pity Francis was not more definite; for our part, we are inclined to award the honour of being the birthplace to the first-named site. As it is, we must not dogmatize.

Shortly after his birth the family moved to a modest house at Penponds,[1] near Camborne.

Young Trevithick was sent to school at Camborne, then a mere village, about a mile distant. Little instruction was to be had there, for the knowledge of the teachers in such schools was extremely limited; indeed few children ever went to school at all, and those who did left at seven or eight years of age to work in the mines. We need hardly wonder then that the boy, although owing to his father's position he enjoyed more advantages than his fellows, did not get beyond the three R's. Further, the master "reported him a disobedient, slow, obstinate, spoiled boy, frequently absent and very inattentive",[2] and no doubt the master was right. The boy had no use for abstract knowledge and, though he showed a certain aptitude for arithmetic, he flouted convention and arrived at results by methods known only to himself. His schooldays were soon over, and it would seem that his father was not more successful than the schoolmaster in moulding and taming this independent son. He was left to wander over the mines, not a matter of difficulty at any time but particularly easy for him on account of his father's connection with them. Experience gained like this gets into the very blood.

During the boyhood of Trevithick there occurred in the mining industry a most important event, destined to have enormous influence over his career. This event was the coming into Cornwall in 1777 of Boulton and Watt, of Birmingham, with

---

[1] The house is still standing and has been restored as far as practicable to the condition it was in at the time, to be preserved as a memorial.

[2] *Life*, I, 52.

PLATE II. RICHARD TREVITHICK AS A BOY
From a miniature in the possession of Capt. R. E. Trevithick

Watt's improved engine, patented in 1769, in which the steam was condensed in a vessel separate from the cylinder. The resultant economy enabled Watt's engine to perform the same work as the existing atmospheric engine with a consumption of between one-third and one-fourth of the fuel. In Cornwall, where coal was so dear and pumping so onerous, the consequence was that no adventurers could afford to do without the engine. At first there was scepticism and opposition, as there always is with anything new. In 1778 Watt wrote to Boulton that the "Infidels of Dolcoath are obliquely enquiring" about the engine. In 1779 he reports a stormy meeting at Wheal Union, where Trevithick senior led the opposition. Watt writes: "During this time I was so confounded with the impudence, ignorance and overbearing manner of the man that I could make no adequate defence, and indeed could scarcely keep my temper which, however, I did perhaps to a fault". By 1780 the adventurers and managers, not least among them Trevithick senior, were so satisfied with the performance of the first engines erected, that there was what amounted to a rush to instal the new engine. Its introduction was so rapid that, in the first six years, twenty-one were set up; the only remaining atmospheric engines ceased work in 1790, although two at least appear to have survived at Dolcoath, for they were altered by Trevithick early in the nineteenth century. During the period 1784 to 1788 eighteen engines were put up; then the pace slackened, none were put up in the years 1789 and 1790 and only five between 1790 and 1796.[1] A great advantage of the new engines was that they increased the depth at which mining could be carried on to more than twice the 50 to 80 fathoms below adit of the old fire engine. There can be no gainsaying that Boulton and Watt were the greatest benefactors that Cornish mining has ever had.

The mode of erection of the new engines was similar to that of the older ones. Boulton and Watt in effect acted as consulting engineers and supplied only a few special parts, such as the valve gear. The adventurers ordered and paid for all the materials, besides paying the wages of the erector, the right of

[1] Dickinson and Jenkins, *James Watt and the Steam Engine*, 1927, p. 328.

appointment of whom was reserved by the patentees. The firm recouped themselves by taking a "premium" or royalty, which they fixed at one-third of the savings in fuel of the engine as compared with an atmospheric engine doing the same work.[1] They modified their practice later, in Cornwall especially, by accepting a fixed rate based on the size of the engine.

It was during this stimulating period of activity in engine building, and in the presence of new men from outside the county, that young Trevithick passed his boyhood. Is it to be wondered at that he preferred such scenes to those of the humdrum school? The time came, however, when it was necessary for him to take up definite employment. His father wished the boy to follow in his footsteps and obviously it would have been a good opening for him, but the young man had other ideas; there could be no question in his case but that it should be engineering. How he spent the intervening years after leaving school we have conjectured above, and in confirmation we have this somewhat meagre statement of Richard Edmonds, who knew him well:[2]

With scarce any schooling, and with no books, he acquired such practical knowledge of steam-engines and mine-machinery that long before he attained his majority he was, to the utter astonishment of his father, appointed engineer to several mines. The father begged the mine-agents from whom the appointment had proceeded to re-consider what they had done, as he was sure his son could not, at so early an age, be qualified for so responsible an office. But having had sufficient proof to the contrary, they merely thanked him for his disinterested advice.

The earliest evidence of such occupation is afforded by the account books, already mentioned, of his father. At Eastern Stray Park Mine in March, 1790, young Trevithick, barely nineteen years of age, is recorded as receiving 30s. a month, but in what capacity is not stated; nor do we know how long previously he had been so engaged. As we see by comparison with figures given above, the wage was a high one. His name continues to figure on the monthly pay sheet for the next two years

[1] *James Watt, loc. cit.* p. 345.
[2] Edmonds, R., *The Land's End District*, 1862, p. 257.

till February, 1792, after which he receives one payment for four days at the same rate of pay.

The period that had just elapsed had been one of depression in the mining industry and attention must have been directed to effecting economies. What more natural than to look to a diminution or remission of the premium payments on the Boulton and Watt engines? With the disappearance of the last atmospheric engine, there was nothing to remind the Cornishmen, or perhaps we should say the younger of them, of the immense savings that had been effected by the introduction of the Watt engine. All they could see was the burden of premium payments that had to be borne; that the payments were considerable may be judged by a calculation made by Boulton in 1787[1] that they amounted to $1\frac{1}{2}$ guineas a ton on the copper and 2 guineas a ton on the tin produced; yet looked at impartially, the payments were but a small part of the cost of production.

Then there were some engineers and adventurers who had doubts as to the validity of Watt's patent and coquetted with the idea of upsetting it. When the worldly wise Boulton informed Watt of this attitude of the Cornishmen towards his beloved patent, the latter became really roused, forgot his usual phlegm and replied in a magnificent outburst:[2]

They charge us with establishing a monopoly, but if a monopoly, it is one by means of which their mines are made more productive than ever they were before. Have we not given over to them two-thirds of the advantages derivable from its use in the saving of fuel and reserved only one-third to ourselves, though even that has been reduced to meet the pressure of the times? They say it is inconvenient for the mining interest to be burdened with the payment of engine dues; just as it is inconvenient for the person who wishes to get at my purse that I should keep my breeches pocket buttoned. It is doubtless also very inconvenient for the man who wishes to get a slice of the squire's land that there should be a law tying it up by an entail. Yet the squire's land has not been so much of his own making as the condensing engine has been of mine....Why don't they petition Parliament to take Sir Francis Bassett's mines from him? He acknowledges he has derived

---

[1] *James Watt, loc. cit.* p. 334.

[2] Letter of 1780, Oct. 31st, cf. *James Watt, loc. cit.* p. 53.

great profits from using our engines which is more than we can say of our invention.... We have no power to compel anyone to erect our engines. What then will Parliament say to any man who comes there to complain of a grievance he can avoid?

It is only too true that the average man looks upon real property as sacrosanct but property in ideas as something to be annexed without payment.

This attitude on the part of the Cornishmen laid them open to entertain any proposals or inventions to take the place of the Watt engine and such were not long in being brought forward. Jonathan Hornblower, in 1781, patented a double cylinder engine—really the first compound engine—but Watt claimed that, condensing in the lower part of the second cylinder as Hornblower did, this was in effect an infringement of the separate condenser. Hornblower had a good deal of trouble with his engine and it was not till 1791 that he was able to put up at Tincroft Mine one that worked satisfactorily. To examine and report on this engine young Trevithick was employed and from his figures it was calculated that the duty was $16\frac{1}{2}$ million, (i.e., of lb. of water raised 1 ft. high by the consumption of 1 bushel of coal), or, the same roughly as that of a Boulton and Watt engine of corresponding size. The fact that this young man of twenty was so employed is convincing testimony that he had attained to a recognized position as an engineer. The original report has fortunately been preserved[1] and we are able to give an illustration (see Plate III).

The note by Giddy on the report refers to a trial made at the same date by Captain Joseph Morcom on a small Boulton and Watt engine at Seal Hole, from which he deduces the duty of the Hornblower as compared with the Watt to be in the proportion of 1·65 : 1.

For the next three or four years we have very fragmentary and one-sided glimpses of what Trevithick was doing. Certainly they were salad days, for we learn much of his exuberant spirits, great stature and bodily strength which led to many a prank, such as his tussle with Captain Hodge, a man as big as himself,

---

[1] Among the *Enys Papers*, Hornblower docket.

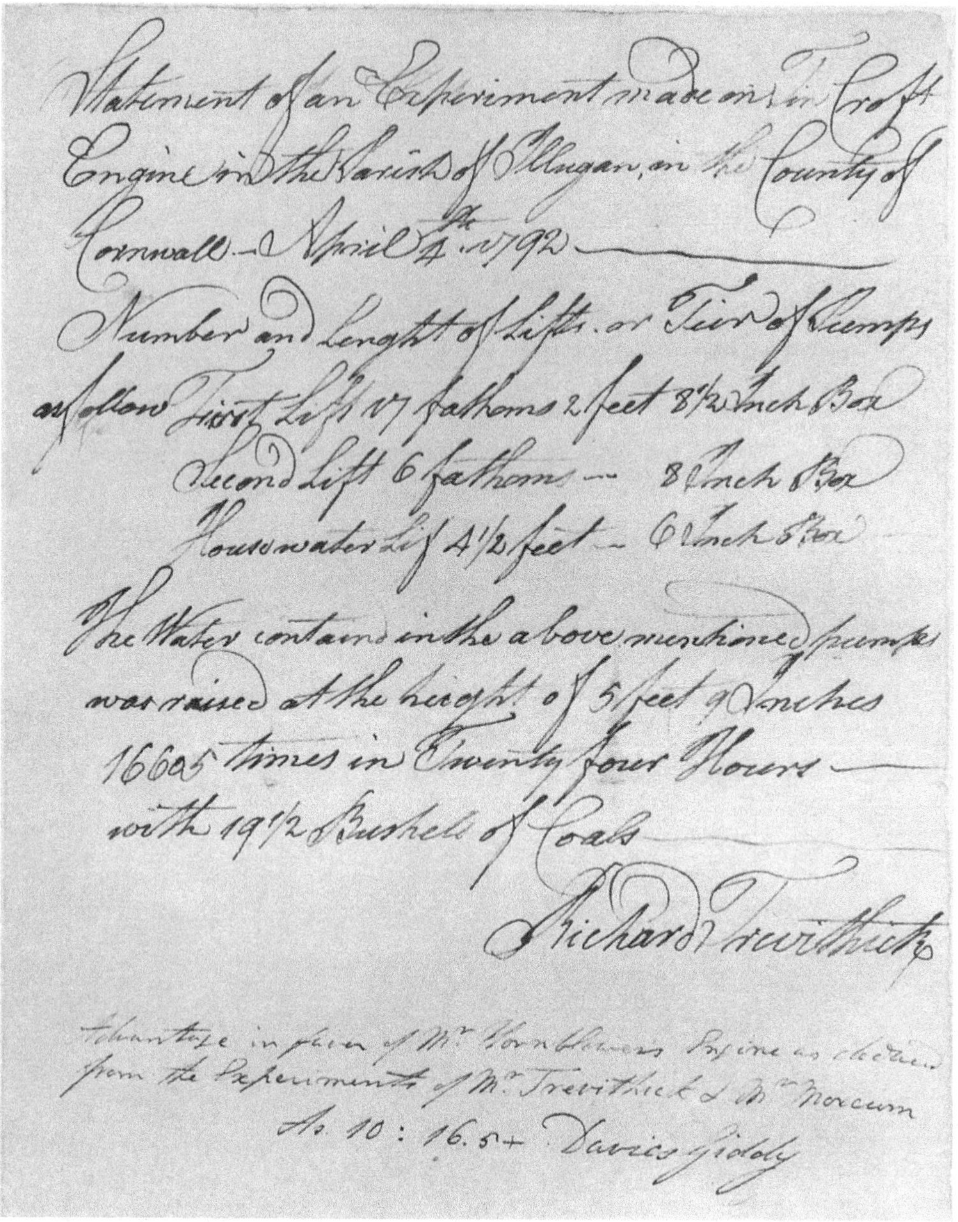

Statement of an Experiment made on the Croft Engine in the Parish of Illugan, in the County of Cornwall – April 4th 1792

Number and Length of Lifts, or Tier of Pumps as follow First Lift 17 fathoms 2 feet 8½ Inch Box
Second Lift 6 fathoms — 8 Inch Box
House water Lift 4½ feet — 6 Inch Box —

The Water contained in the above mentioned pumps was raised at the height of 5 feet 9 Inches 1665 times in Twenty four Hours with 19½ Bushels of Coals

Richard Trevithick

Advantage in favor of Mr Hornblowers Engine as deduced from the Experiments of Mr Trevithick & Mr Norcum As 10 : 16.5 + Davies Giddy

PLATE III. HOLOGRAPH REPORT BY TREVITHICK ON
HORNBLOWER'S COMPOUND ENGINE, 1792
From the *Enys Papers*

after one of the monthly account dinners at Dolcoath; Trevithick seized Hodge by the waist, turned him head downwards,
and stamped the imprint of his boots on the ceiling of the room!
Another of his feats was to write his name on a beam 6 ft. from
the floor with half a hundredweight hanging on his thumb. In
competition with other kindred spirits, he is recorded as having
lifted a 9 in. pump barrel "weighing 7 or 8 cwt.", and not only
so but to have hoisted it on to his shoulder and borne it off the
field. At Crane Mine, some "young men standing at the door
of the smiths' shop tried to throw a sledge [hammer] against
the wall of the engine-house . . .; but Captain Dick happening
to come by, threw the sledge across the yard and over the roof of
the engine-house".[1]

Hugh Hunter, foreman carpenter in Cook's Kitchen Mine, in
1869 related the story of Trevithick lifting the mandril[1] in the
smith's shop there. He had

often seen Captain Trevithick lift the mandril, and hundreds of people
used to come and see him do it. He used to put a bar of iron inside the
mandril and fasten another bar to it so that he could get a good hold.
A strong stool was placed on each side of the mandril, upon which he
would stand with the mandril between his legs, and would lift it off
the ground. He was an uncommon quick-spirited man and the strongest
ever known. The weight of the mandril was ten hundredweight.[2]

Another story, obtained at first hand, is that Trevithick would
climb the pit-head gear at Cook's Kitchen Mine and there engage in what we may call eurhythmics with a sledge-hammer.
"He did it for exercise and to steady his hand and his foot!!"
Stories of his immense physical strength were current in Cornwall forty years after his death, so greatly had they impressed
themselves on people's minds. He was skilled in the Cornish
sport of wrestling, and his son tells us that[3]

His excusable vanity was gratified by a request from a member of the
College of Surgeons that he would show them his strong frame; and
their telling him that they had never seen muscle so finely developed,
interested his Cornish friends, who delighted in physical strength.

[1] *Life*, I, 55.
[2] A smith's mandril is a thick cast-iron conical pipe.
[3] *Life*, I, 54.

To return to his serious occupations, we find Trevithick at Wheal Treasury[1] in July, 1795, engaged in "moving and erecting the eastern engine 21*l.*" He was paid "for one month's attendance on the engine 1*l.* 11*s.* 6*d.*" The item "for two months' saving respecting the eastern engine 18*l.*" occurs but it does not appear that it was paid to Trevithick. Another entry is: "To two months' savings respecting Mr Bull's engine 43*l.* 12*s.*" Bull likewise receives a monthly payment of 1*l.* 11*s.* 6*d.* These entries continue to appear till May, 1796, when all engine costs ceased. One of the last entries is: "Capt. Trevithick for his attendance 1*l.* 1*s.*" The mine closed down in December, 1797. It is not easy to explain these entries. Boulton and Watt had erected a 36 in. single-acting beam pumping engine at Wheal Treasury in 1780 but by this time it had been moved to Wheal Gons. Two engines are distinctly specified in the entries: one erected by Trevithick and one by Bull. Possibly they took the place of the Watt engine and it may be that it was at Wheal Treasury that Bull and Trevithick first met one another—a meeting that led to an association of several years' duration.

Little is known of Edward Bull. We hear of him first in 1779 as engine-man at Bedworth Colliery, Warwickshire, the engine for which was erected by Boulton and Watt. He was in 1781 sent into Cornwall by the firm as an erector under an agreement. He was in their employment as late as 1789 at Wheal Virgin, when Wilson, under date September 22nd, mentions[2] "Bull's scheme"—no doubt some fancied improvement in the engine. By 1791 Bull must have left their service and become sufficiently well known to be engaged elsewhere, for James Watt writes:[3] "We have a long letter from Bull wanting us to permit him to erect our engines to fulfil his contracts, but not to be subject to you or Mr M." (i.e. Murdock).

As we have indicated above, such a course was definitely op-

---

[1] Trevithick senior's Account Books.

[2] *Boulton and Watt Letters*, 1780–1803, being the correspondence with the firm of Thomas Wilson, their agent in Cornwall, in possession of the Royal Cornwall Polytechnic Society.

[3] *B. and W. Letters*. J. W. to T. Wilson, 1791, Sept. 26th.

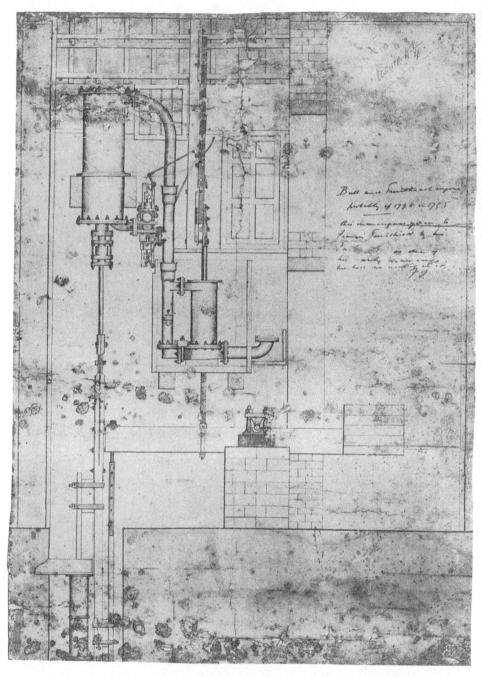

PLATE IV. ORIGINAL DRAWING BY TREVITHICK
OF BULL'S ENGINE, *c.* 1792

In the possession of Lord Falmouth

posed to the firm's policy and was naturally refused. Watt's comment on the decision was:[1]

In respect to Bull the less we have to do with him the better. If he applies to you on our terms & brings respectable persons as principals you will fix the premium with him & take his order for the size of the engine but we will not be directed how to make it.

Had we agreed to let him make one of our Engines in such manner as he pleased, he would have made a bad thing & we should have had our share of the disgrace. As it now stands, his inventions must depend upon their own merit & unless he becomes more knowing than he has been hitherto, the merit will decide in our favour.

Evidently Bull had not yet designed the distinctive inverted type of engine with which his name is associated, for it was not till the following year that he put up at Balcoath his first engine of this type. Its principal feature was that the pump rod was directly coupled to the piston rod and the upper part of the cylinder was always open to the condenser and consequently infringed Watt's patent. Bull does not appear to have claimed to have invented it; indeed the arrangement was not new, for Watt had schemed it in 1765, and John Wilkinson made one, known as the "Topsey-Turvey" engine, in 1776. Watt, however, put it aside in favour of the beam engine. Bull's engine certainly looked very different from the beam engine and perhaps was cheaper to make; at any rate its merits have enabled it to survive to the present day.

Young Trevithick made a drawing of one of Bull's engines, possibly the one at Balcoath. Fortunately it has been preserved, because it was given by Trevithick to his son Francis as a creditable piece of draughtsmanship; that it is so an inspection[2] (see Plate IV) will reveal. The chief interest of the drawing is the light that it throws on the association of Trevithick with Bull, for the former would hardly have made such a drawing otherwise. In all, Bull put up ten engines, and it is probable that Trevithick had a hand in most of them.

[1] *B. and W. Letters.* J. W. to T. Wilson, 1791, Oct. 10th.
[2] *Life,* i, 59. Drawing in the possession of Lord Falmouth, and reproduced by his kind permission.

There is only one conclusion to be drawn, and that is that Bull, who was a poor man, was backed up financially by the Cornish faction who sought to invalidate Watt's patent and thought they saw in this engine a means of bringing the matter[1] to trial, which is in fact what happened. We have gone into Bull's career somewhat fully to stress the point that by this association Trevithick ranged himself definitely, as his father had done before him, on the side of the opponents of Boulton and Watt.

The firm proceeded against Bull, and *pendente lite* in 1794, March 22nd, the Lord Chancellor granted an injunction to restrain him "from erecting any more engines upon Mr Watt's plan—and from compleating those he has in hand".[2] Bull then asked the Court to be allowed to finish the engines preparing for Ding Dong and Hallamannin Mines, but this, too, was refused.

Following this up, Mr Weston wrote:[3]

...the Lord Chancellor has this Day granted an Injunction against using the engine at Ding Dong Mine. Mr Trevithick Junr. could not be included as he has not been made a party to the Suit, but if you will find me his *Christian* name, I will take care to prevent his future intermeddling; and further if you can send me an affidavit that Trevithick is Bull's Partner or his known Agent, I will apply to the Court to grant an *Attachment* against Bull for his contempt.

Again, on June 22nd, Weston wrote:

I hope you will be able to discover how far Richard Trevithick has been concerned in erecting the Engines at Ding Dong & Poldice, and also how far he is connected with Bull.

Trevithick was now assuming almost as much importance in the eyes of Boulton and Watt as regards infringement as was Bull, and Weston wrote on June 29th:

...You said in one of your Letters lately that *Trevithick had got the Engine* [i.e. Ding Dong] *to work*, saying that he was not subject to or under the Injunction or to that effect:—if this can be sworn to, let it be done.

---

[1] *James Watt*, *loc. cit.* p. 309.

[2] *B. and W. Letters.* A. and T. Weston to T. Wilson, 1794, March 22nd.

[3] *B. and W. Letters.* T. Weston to B. and W., 1795, June 11th.

In spite of all efforts on the part of Wilson to get evidence that Bull and Trevithick were partners, nothing but hearsay was forthcoming, for the sufficient reason, in the authors' opinion, that no partnership existed. Boulton and Watt, being unable to prove partnership, had no other course open to them than to get an injunction against Trevithick separately. The effect of the injunction at Ding Dong was that Trevithick altered the engine to an atmospheric; later on he altered it again to a Watt type and premiums were paid upon it to Boulton and Watt.

We may digress here to give the octogenarian reminiscences in 1869 of Mrs Dennis,[1] who knew Trevithick at this time, particularly as they throw a sidelight on his personal character:

Mrs Dennis recollected Mr Trevithick at Ding Dong about 1797 fixing his new plan of pumps there, and at Wheal Malkin and Wheal Providence, adjoining mines. Her parents lived at Madron, near these mines, and for two or three years Mr Trevithick came frequently to superintend the mine-work, staying at their house a few days, or a week at a time. He was a great favourite, full of fun and good-humour, and a good story-teller. She had to be up at four in the morning to get Mr Trevithick's breakfast ready, and he never came to the house again until dark. In the middle of the day a person came from the mines to fetch his dinner; he was never particular what it was. Sometimes, when we were all sitting together talking, he would jump up, and before anyone had time to say a word, he was right away to the mine.

This is first-hand testimony to his kindly disposition, tireless energy and sudden impetuosity.

An incident that created some bad feeling now occurred, and this was the use of threatening language—it is not clear whether it was on the part of Trevithick—to Simon Vivian, one of Boulton and Watt's erectors, and to William Murdock. It will lead to a better understanding of the situation if it is realized that James Watt junior, who had come back from France only in 1794 after his travels, had entered the firm. Incidentally this will clear away the confusion that has arisen in attributing to the father actions of the son. James Watt and Matthew Boulton had treated the Cornishmen with much consideration by abating or

[1] *Life*, I, 72.

remitting premiums; for example, up to 1794, Poldice adventurers had been let off payment of premiums amounting to £13,000.[1] Of course the Cornishmen did not want to pay anything at all and adopted any pretext to avoid payment. James Watt junior and Matthew Robinson Boulton, who were now assuming the direction of affairs of Boulton and Watt, were cast in a different mould to their fathers, and seem to have been so antagonized by the Cornishmen's attitude that they determined to have their "pound of flesh" out of them. James Watt junior, especially, was a hot-head and given to immoderate language such as his father would not have used, as we see in the following letter:[2]

P.S. ...You see that both Bull and Trevithick give you all the lies, but particularly to Simon Vivian. Tell him however not to mind their damned impertinence. They have foresworn themselves out and out and I hope before long, we shall give a good account of the rascals. I hope Vivian will exert himself to get more information about the Partnership between Bull and Trevithick. They are both done for if we can prove that point. S. Vivian at all events shall be properly provided for, but he may be of use in Cornwall yet.

The next day James Watt junior wrote:

As to our good friend Vivian...it is now his interest as much as ours to procure additional proofs of the partnership between Bull and Trevithick. It is better he should be actually turned out of employ by the Captains &c., than that he should leave them voluntarily...if he cannot without fear of his life, remain in the County, let him come here with his wife and family in God's name.

To secure him from intimidation, Vivian and his family were brought to Birmingham, and we shall hear of him there later.

There is a hint in the next letter we quote that the idea of going abroad had taken root in Trevithick's mind:[3]

I apprehend that what Trevithick Jr. says about his going to the East Indies is merely a sham to prevent his being served with the Injunction, hope therefore you will not fail to press it being served.

---

[1] *James Watt*, loc. cit. p. 334.

[2] *B. and W. Letters.* J. W. junior to T. Wilson, 1795, July 20th.

[3] *B. and W. Letters.* J. W. junior to T. Wilson, 1795, Oct. 13th.

By the end of the year the injunction was still unserved upon Trevithick. No doubt he knew about it, and although he kept out of the way for the time being yet it must have had an effect in bringing him round gradually to the conclusion to try to make terms with Boulton and Watt.

In May, 1796, Mr Kevil of St Agnes wished to give Trevithick the job of erecting a Bull engine there, but again his proposal was rejected in a long letter from Watt:[1]

We shall agree to Mr Kevil's erecting one of Bull's engines on the terms you mention but as we are situated respecting Trevithick Junr. *we cannot* consent to his being employ'd, nor do we think it Mr K's interest to have anything to do with him....As Trevithick Junr. is Df^d to some of our bills and has never made answer nor suffered the injunction to be served upon him we should injure ourselves very materially were we to countenance his being employed in any way unless he first makes his peace by a full and fair confession of the facts in his power to prove, nor can we in common justice thus give him a preference over Mr Murdock who has served us faithfully and is perfectly qualified to do the Engine Justice. If we have any right to the business we have a right to employ our agents in the most essential part, the erection. We hope these reasons will satisfy Mr Kevil, whose candour cannot press us to a thing which would be injurious to us.

Probably this was the last straw, for very shortly after a letter came, ostensibly on his behalf, to Boulton and Watt, from Captain Thos. Gundry, a man who was held in high esteem by the firm; the letter is as follows:[2]

Gentlemen, I have taken the liberty to trouble you with this by desire of Richd Trevithick J^r who have been for some time past employed by Edwd Bull in mechanism. He desires not to continue in opposition to you, and is ready to give up everything in this county and be under your direction. If you should employ him, you will certainly find him possessed of good abilitys in mechanics, natural as well as acquired, and is of an honest and peaceable disposition. He would be glad to serve you either in Cornwall or Soho, *the latter place in particular*. If this step is taken I think the opposition in Cornwall would to a great measure subside. I would esteem it a peculiar favour

---

[1] *B. and W. Letters.* J. W., Soho, to T. Wilson, 1796, May 9th.

[2] *James Watt, loc. cit.* p. 310.

if you would take the matter into consideration and don't doubt but that any favour conferred will be gratefully acknowledged by him as well as yr. mo. obed^t Servt

THOS. GUNDRY.

Goldsithney, 5th July, 1796.

The letter was referred to Wilson, who spoke up for Trevithick, which caused the firm to modify their attitude to the limited extent shown in this long-winded letter:[1]

My father and Mr Weston concurring with us all, in the propriety of disengaging young Trevithick from the confederacy, to which your good character of him has not a little contributed, we have determined to withdraw our suit against him, and if the writs are not already served, desire you will call them in, if they are they must be proceeded upon. The inclosed letter to him will sufficiently explain our sentiments upon that head, but with respect to engaging him in our service we shall be entirely guided by the inclinations and advice of Mr Murdock, whom we feel ourselves bound to consult and to be governed in this matter. We inclose a letter to him open for your perusal but request that both it and the one to Trevithick may be sealed before they are delivered. If William approves of our employing him, we think it will be that he should carry the letter himself; if not you will send it to him.

Should Mr Murdock approve of our employing him, as you say he does, it will remain to be considered whether he should be employed here or in Cornwall. We think he might be useful in Cornwall to destroy the prejudice against us, which [he] himself has helped to circulate and in that case we must leave it to you to fix his wages; but if it [is] thought preferable to employ him here, *the utmost* wages we can offer would be a guinea per week[2] and we could not take him into our works upon a less engagement than for three years, but we should prefer five. If old Trevithick will give security for his son, so much the better, but we should not absolutely insist upon that, considering the good character of the young Man, provided he enters into regular articles of service, which indeed must be insisted upon in every case, whether he remains in Cornwall or comes here. As soon as that is determined on we shall send you a Copy of Articles. If it is determined to employ him in Cornwall of course take the first opportunity

---

[1] *B. and W. Letters.* J. W. junior to T. Wilson, 1796, July 19th.

[2] "It would perhaps be better to offer less for the first two years, say 19/- and 20/-", is added in the margin.

of conversing with him upon the subject and of informing us of his expectations &c....

[P.S.] We particularly desire to know whether Trevithick is a good workman, or whether he has only been accustomed to superintend work, also what are his other qualifications.

Although the wages offered seem extremely small, they were the standard rate that the firm was paying to its erectors, e.g. Murdock got only a guinea a week when he started work, but we can hardly imagine the wages, coupled especially with the prospect of being tied down to a three years' engagement, could have had the slightest attraction for Trevithick. We imagine rather that he used some strong language, and it is not surprising that he did not trouble to reply to the letter that was sent him. Besides, Murdock clearly did not want him, so that the negotiations were broken off in the following letter:[1]

As Mr Murdock has no wish for Trevithick to be employed, and he is himself become indifferent about it and has not had the good manners to answer our letter, we shall give up the idea of taking him into our Service, more particularly as we see no reason for augmenting the wages mentioned in our last, which are the utmost we give to putters up of Engines, the only capacity in which he might be useful to us. We are however not inclined to renew hostilities against him notwithstanding his setting us at defiance, unless he commits some flagrant act of contempt in which case the Injunction *must and shall be served*, even if one of us comes down in person to do it.

It appears to the authors that in making this move towards *rapprochement* with Boulton and Watt, Trevithick was acting as the spearhead of the Western Mine captains with the intention of getting a footing in their opponents' camp. There is no doubt that a system of espionage was carried on by both sides. The injunction still remained out and sure enough at last it was served on Trevithick and at Birmingham of all places. He, Edward Bull and Andrew Vivian had gone there together, but on what business we do not know except that it was ostensibly on mine business, because their expenses were paid by Wheal Treasury

---

[1] *B. and W. Letters.* J. W. junior to T. Wilson, 1796, Aug. 9th.

as its Cost Book shows. We must give the incident in Matthew Boulton's own words:[1]

Woodward found Trevithick and his Friends at a publick House facing my Manufactory & delivered to him ye Injunction which he rec[d] with much surprise particularly as he thought nobody knew him. He seemed much aggitated and vexed; however he afterwards went wth Bull and Andrew Vivian to dine with Simon at the Foundry where he found our Men fireing of Canonons [*sic*] & rejoicing at our Victory w[ch] took away his appetite from his dinner. Andrew was admitted to see the Foundry & Manufact[y] but not the others. It is rather curious that although ye Injunction could not be served in Cornwall T: should run into the Lyons Mouth & afterwards go to Dine wth the man that they had Banished from Cornwall. They afterwards went to Colebrook Dale but know not their reception.

The "victory" that caused all the rejoicing was the news that the action Boulton and Watt *v*. Hornblower and Maberley had gone in favour of the firm. It was a glorious triumph for Boulton and Watt and everyone gloated over it.

In another letter at the same time from John Southern we have this delicious touch:

Bull took his dinner quietly; but Trevithick walked backwards and forwards in the house like a mad man, and firmly resisted all temptation to dinner; till the smell of a hot pye overcame his powers, on which he set to and did pretty handsomely, but in such a manner as shewed him not quiet in mind.

Trevithick continued, however, to offer his services to adventurers to erect engines. The next offer of which we learn was for the United Mines:[2]

We learn from R. Mitchel that Bull & Trevithick are constant attendants at the United Mines & that they have offered proposals for the erection of their new Engine. We wish their motions to be observed but not disturbed; by giving him length of rope we have no doubt but they will get entangled & the injunction may be enforced when they least expect it.

[1] *B. and W. Letters.* M. Boulton to T. Wilson, 1796, Dec. 26th.
[2] *B. and W. Letters.* M. R. Boulton to T. Wilson, 1797, May 19th.

James Watt junior expressed himself on the subject, in his usual forcible manner, in the following long-winded letter:[1]

We have not felt one moment's hesitation as to the propriety of totally rejecting every idea of connecting ourselves with Trevithick and of withholding our consent for the erection of our Engines by him or anyone else upon the United Mines until all the Arrears are liquidated and we are assured that they will continue to pay in future. You will therefore send for Trevithick and inform him VERBALLY that his proposal is totally inadmissible and that no consideration of present emolument, were it ever so well secured, would induce us to permit *any* of our principles to be applied until we have received compleat indemnification for the past and security for the future. As to Bull's threats we know the Rascal is too well-tied up by the Injunction to dare to move a step, & the moment he breaks down the fence, we shall take care to have the scoundrel secured and the Adventurers put under a similar restriction. This case is exactly similar to that of Poldice and what happened there will almost probably be acted over again. This is all you have to say from us to Trevithick. This repentance for past offences comes too late and his promises of future good conduct meet here with little credit. If his own conscience is insufficient to induce him to act with common honesty, we shall not bribe him so to do—Let him therefore take his swing.

We next find Trevithick changing his ground and proposing to erect a Watt engine, as witness the following:[2]

Trevithick's present proposition appears to us to bear a more likely aspect than the preceding one, because it is certainly his interest to pay our Premiums & erect one of our Engines in preference to any other. However it is necessary for us to be extremely guarded in our proceedings with him and we have therefore resolved that you should acquaint him verbally from us that we had received his request transmitted by you, that it appeared strange, if he meant fairly that he did not himself write to us upon the subject; that after the repeated proofs he had given of an inimical disposition towards us, we must naturally receive with caution and distrust any proposition coming from him unless seconded by persons of known respectability and that until we are convinced of the sincerity of his intentions, it would be absurd in us to enter into any negociation. You may add that upon a satisfactory letter being written to us by himself upon the subject, seconded by one from

[1] *B. and W. Letters.* J. W. junior to T. Wilson, 1797, June 2nd.
[2] *B. and W. Letters.* J. W. junior to T. Wilson, 1797, June 27th.

his father, offering security for him, his application would probably meet with attention from us. You may add that a letter from Mr Kevill or any other of our Cornish friends in his favour would have considerable influence upon our determination.

James Watt, more open-minded than his son, adds on the last sheet: "If T. means fairly he should be fairly dealt with".

Still another move on the part of Trevithick was that of a visit to Soho to try to arrange specially reduced terms for Wheal Treasury:[1]

Messrs Vivian & Trevithick have left us. We could not agree upon terms. T. offered £800 for the arrears of Wl Treasury old adventurers & ½ premium (55 for a 63) for some engines he wanted to build also for Wheal Crenver & others in which Harris is concerned. We insisted upon 2/3rd for the new erections & the full for litigated arrears. V. offered £30 per month for all the engines on Wheal Treasury to commence with the entry of the new adventurers, we insisted upon 2/3rd for the whole of their time according to the engines at any time actually working. He offered half premiums for Wheal Unity (or I am not sure whether only £30 pr month) and in the same ratio for some others. We refused all the offers & rested upon what has been said. We believe we can recover the whole of Wl. T. upon application to Chancery which however we are loth to do.

They talked much of the good performance of T's engine at Dingdong which they say does 23 million p^r bushel & that ours of the same size do not do above 16. This we cannot entirely credit.

We are not sorry to be off with T. as we do not like him to interfere with Mr M. whom we should however have taken care of in some other way, when such interference was likely to be hurtful. In the present case he could not have been affected as the engines were already undertaken by T. on his own account, he being to find engines, men & coals, for a monthly sum. He said if he could agree with us he had offers from some other mines which had our engines to take them by the month & by putting them in better order than they are he could get money by them after paying our premium.

This solicitude of Watt for Murdock is most admirable. This letter was followed by a suggestion from M. R. Boulton that Wilson himself might work in conjunction with Trevithick.

[1] *B. and W. Letters.* J. W. junior to T. Wilson, 1797, Oct. 5th.

Further light on the position is thrown, and the real reasons that the young partners in Boulton and Watt had for employing Trevithick are given, by the following letter:[1]

> ...It occurs to us that should Murdock from disgust wish to leave the county and come to reside at the foundry such an agreement, provided there are no legal exceptions to it, would facilitate the settlement of our contested claims, and be a means of avoiding the unpleasant task of bargaining in settling future agreements. Old Trevithick's death may probably be a reason for the son's desiring to renew his engagement with us and may have put him in possession of property to guarantee the stipulations on his part...the conduct of the negociations especially of Vivian has been open and candid, and procured his colleague an invitation to dinner.

All along the impression that the correspondence gives is that Boulton and Watt looked upon Trevithick as a formidable opponent whom they would only too well have liked to have gained over to their interest; there was a constant swing of the pendulum from one position to the other, both sides playing fast and loose with each other.

The next incident of importance which the correspondence tells us is a proposal by Trevithick to erect an engine at Prince William Henry Mine (Wheal Chance or Roskear). This is what is referred to in the following letter[2] from Messrs Fox of Truro who were friends of Boulton and Watt in Cornwall. Its importance resides in the categorical denial by Trevithick that he was in partnership with Bull:

> A report having been circulated hereaway (and probably may have reached you) that Edw[d] Bull is a partner with R[d] Trevithick in one or more steam engines that he is about to erect by your permission, we have had some conversation with R. T. on the subject who seems to be much hurt by a report which he asserts to be invidious and totally void of foundation—this we are disposed to believe and at his request take freedom of intimating this much to you....

[1] *B. and W. Colln.* Letter, M. R. Boulton, Soho, to J. W. junior, 1797, Sept.

[2] *B. and W. Colln.* Letter, Messrs G. Fox, Perran Wharf, to B. and W. 1797, November 29th.

The proposal is further discussed at great length in the following letter:[1]

Trevithick's conduct is such a medley composition that we shall not trouble ourselves to unravel it & it is of little signification whether we are to add to his other transgressions a connexion with Bull. Without any confirmation of this report you know we have sufficient grounds for mistrusting his designs & feigned submission. His late equivocation in intimating to Messrs Foxes that he had our permission to erect Engines is of itself ample proof of the little reliance to be placed upon his assertions. Why Trevithick should expect our notification of terms which he had rejected with disdain is not less inconsistent with common sense than the whole of his chequered conduct since he resolved upon his concilatory plan. He must or ought to be sensible from the manner in which he concluded the negotiation at Soho that we should not renew the terms then refused. When we parted he fully gave us to understand that rather than accede to our demands of the 2/3rd premm he should continue to erect his atmospherl Engines & most strenuously asserted that he could work them cheaper than ours under the terms required by us.

You have now another instance of his duplicity that while he was making protestations of this tenor to us he had secretly resolved to convert the Engine to our principles & which intention you see he has carried into effect in defiance of law & his solemn assurances to the contrary. It would be in vain to make any agreement with a Man of his stamp without security in hand for its performance & we have therefore only one offer to make viz. to pay the premium of Ding Dong & the other engines where our principles are employed in one sum calculated after the same ratio as the sum proposed to the Advenrs of Penandra.

We now find a letter from Trevithick himself and the earliest one known to the authors to be preserved; consequently we give it in full:[2]

Redruth, Decr 16, 1797.

Sir,

I Received your Letter last night stateing Mr Boultons & Watts demand on my Engines, as to ding dong Engine saveings I have no objection to pay according to your Statement of £800 for a 63 Inch Cylinder. Ding Dong is 28 Inches which I believe gives a sum of abt

[1] *B. and W. Letters.* M. R. Boulton to T. Wilson, 1797, Dec. 6th.
[2] *B. and W. Letters.* R. T. to T. Wilson.

£150 which sum I wo'd be bound to pay in monthly payment sho'd
the Engine Continue to work Single which I believe will be but a very
short time as every prepareation is already made for the purpose of
working double. Now if its agreeable I will pay you the monthly sum
above stated untill the Engine is turnd double which I expect will be in
the coarse of two or three month and then enter in to bonds for the
remainder of the time for double the saveings above stated—

The terms for St Agniss engine I think is somewhat hard as there is
a certainty that engine cannot be ready to work before the end of six
or eight months and probably unseen obstacles may arise that may
prevent her working much longer as the Resolutions of St Agniss
Mines is not much to be depended on which wo'd be a great Hardship
on my side sho'd it so happen. I have by no means any objection
against agreeing after the same rate for that as I have for ding dong
from the time she shall be set at work which I think is the properest
time to agree as no dependance can be placed on Mines where the
Adv^{rs} is so often in the vice-wardens court. An answer wo'd very
much oblidge

your very Hble Sevn^t

RICH^D TREVITHICK.

James Watt junior now asked Wilson to get proof of what
alteration Trevithick had made on Ding Dong engine and on
December 23rd sent the following draft for Wilson to write in
reply to Trevithick's letter of the 16th:[1]

I have communicated your letter of the 16th to Messrs B. & W. who
desire me to say that as all the preparations are made for converting
Ding-dong Engine into a D'ble one and that the change will take place
in the course of two or three months; they have calculated the Premium upon the supposition of its working two Months single and the
rest of the Term double, which makes the whole sum £310. For this
sum payable in Monthly Instalments they insist upon having your
Bond before further alteration is made in the Engine and I therefore
request that you will appoint an early day for giving me the meeting
to draw it up. I am further desired by Messrs B. & W. to intimate to
you that you are not to proceed to the erection of any other Engine
upon their principles either at St. Agnis or elsewhere until you have
settled the terms with me according to the aforesaid rate & given your
Bond in each case for the amount. They therefore expect that before
you proceed to contract with the Mines to alter any Engine, or to erect

[1] *B. and W. Letters.* J. W. junior to T. Wilson, 1797, Dec. 23rd.

D & T   3

anew any Engine upon their principles, that you shall in each individual case apply for their licence. B. & W. cannot enter into any circumstances attending this or that mine, they consider the licence as granted to you & not to the mines & it is your duty to secure yourself against the adventurers.

The high and mighty tone of this letter must have been the reverse of soothing to Trevithick, and Watt junior's instructions to Wilson as to how Trevithick was to be treated were couched in similar language, as witness the following:[1]

We have your favour of 30th & 31st ulto and having considered the circumstances you mention relative to Trevithick's working single at Dingdong for some time to come, we think upon the whole it will be best to make out the Bond upon that supposition now, and have a fresh bond when it is intended to be made double. The Lump Sum for working single is £158 which if convenient to Trevithick, may be paid at once and there will be no occasion for a bond.... We observe that Trevithick has concluded his Agreement for the St Agnes Engines. As soon as the Dingdong business is concluded and we understand that a formal application is made by him for leave to work them upon our principle, we shall state the sum for which he is to give bond in which we shall make allowance for the time necessary to get them to work.

Apparently by January 15th Wilson had settled with Trevithick and he was at liberty to proceed with the St Agnes engine. On March 8th Trevithick sends bills to the amount of £43 odd in payment on account. In a letter of July 16th Trevithick reported that the Wheal Abraham adventurers had not authorized payment of the premium on their engine. There are letters on September 4th and 24th from Trevithick to Wilson regarding the premiums on Wheal Leeds, and an interview that Trevithick, accompanied by Captain Hodge, had with Wilson on November 27th shows that they settled payment, not, however, to the satisfaction of James Watt junior, who wrote:[2]

We think you have shewn rather too much lenity to Trevithick and his colleague Hodge. Our negotiation was understood to be with

[1] *B. and W. Letters.* J. W. junior to T. Wilson, 1798, Jan. 3rd.
[2] *B. and W. Letters.* J. W., but in the handwriting of J. W. junior, to T. Wilson, 1798, Nov. 30th.

Trevithick and by no means with the Adven^rs of Wh^l Leeds or their representative Hodge. Trevithick was acquainted with our terms before the Engine was or ought to have been erected and he had therefore sufficient grounds for his agreem^t with the Adven^rs. From the many repeated instances of his duplicity we are rather incredulous to his pretence for a modification on the arrears. We are rather inclined to think that the positive assurance of an attachment being the consequence of his transgression would have induced him to comply with the original terms.

Apparently payments were kept up for we find Captain Richard Hodge on May 6th, 1799, transmitting £123 on account in payment of Wheal Leeds engine.

We cannot resist breaking off here to quote from James Watt junior's letter of triumph dated 1799, January 25th, on receipt of the news of the verdict in the appeal Boulton and Watt *v.* Hornblower and Maberley:

Send forth your Trumpeters and let it be proclaimed in Judah that the Great Nineveh has fallen; let the Land be cloathed in sackcloth and in Ashes! Tell it in Gath, and speak it in the streets of Ascalon. Maberley and all his host are put to flight!

The land that was to "be cloathed in sackcloth and in Ashes" was no doubt Cornwall.

We have now arrived at a turning-point in the history of mining in Cornwall, and one that had considerable influence on Trevithick's career. This was the expiration in June, 1800, of the famous patent of Watt for the separate condenser; consequently any one was now at liberty to use it and every engineer was free to develop his ideas without feeling that the shadow of litigation was over him. The firm of Boulton and Watt was known no more in Cornwall except as collectors of money overdue for premiums. M. R. Boulton was sent into the county on this business and he it is who is referred to in the letter from Trevithick on that matter which we reproduce below. A letter in 1801, September 7th, shows that Trevithick had not then fully paid up all that was owing.

We have had only one side of the story in this correspondence, however, so that we must not judge Trevithick on the evidence given, but we can say he did not deserve all James Watt junior's violent strictures. He was doing his best for his employers; he was no kid-glove antagonist and was ready to take every advantage that he could of those to whom he was opposed.

We have, however, proceeded at too rapid a rate, for we have omitted mention in the proper sequence of three events in Trevithick's life that had a most marked influence on his career. To take them in order, the first was the beginning of the staunch friendship of Davies Giddy (1767–1840), a man of much Trevithick's own age but whose career was entirely different, for his mathematical gifts had been fostered in the Oxford schools. He rose to great eminence and at length came to preside over the Royal Society. He took the surname of Gilbert in 1816 on inheriting property from his father-in-law, whose daughter Anne he had married in 1808, but he was always Giddy to Trevithick and will be so referred to in these pages. Giddy's account,[1] written in his latter years, is

About the year 1796 I remember learning from Mr Jonathan Hornblower that a Tall & strong young man had made his appearance among Engineers, and that on more than one occasion he had threatened some people who contradicted him, to fling them into the Engine Shaft.

In the latter part of November of that year I was called to London as a Witness in a Steam Engine Case between Messrs Bolton & Watt & Maberley. There I believe that I first saw Mr Richard Trevithick, Junr., and certainly there I first became acquainted with him.

Our correspondence commenced soon afterwards, and he was very frequently in the habit of calling at Tredrea to ask my opinion on various Projects that occur'd to his mind—some of them very ingenious, and others so wild as not to rest on any foundation at all.

We have been unable to confirm the statement that Trevithick was in London on the date named, but that is a small matter. The important fact is that throughout his life Trevithick consulted Giddy on theoretical questions in mechanics and never failed to obtain illuminating help.

[1] *Enys Papers.* Letter, Davies Gilbert to J. S. Enys, 1839, April 29th, quoted in the *Life*, i, 62.

The other important events were domestic: the death of his father and his marriage. His father died on August 1st, 1797, at Penponds, and was buried in Camborne churchyard. As the outcome the young man's pecuniary position improved, and doubtless that was why he now felt able to marry the girl with whom he had fallen in love—Jane Harvey, daughter of John Harvey of Hayle Foundry. The marriage took place on November 7th at St Erth Parish Church. They were a fine looking couple. The bride was tall with fair complexion and brown hair. She was a woman of great strength of character and of remarkable courage, attributes destined to be severely taxed in the life before her. Marriage with Trevithick was no bed of roses, no ivy-clinging with such a whirlwind was possible. She stuck to him through thick and thin and, as far as we know, never regretted their union. Trevithick is described by his son as "6 feet 2 inches high, broad shouldered, well-shaped massive head, blue eyes, with a winning mouth somewhat large, but having an indefinable expression of kindness and firmness". Richard Edmonds junior says of him:

He was unassuming, gentle, and pleasing in his manners; his conversation was interesting, instructive, and agreeable, and he possessed great facility in expressing himself clearly on all subjects....His dress was plain and neat, and his general appearance such that a stranger passing him in the street would have taken him for some distinguished person.

The young couple lived for the first nine months at Moreton House, Redruth, and then moved to Camborne, where they remained for ten years. An amusing story is told of Trevithick's carelessness on removal: he forgot to give back the key of the house because he had left it in an old coat pocket. His forgetfulness cost him a year's rent.[1]

He now settled down as an engineer and became more and more in request. Some of his account books have been preserved (see p. 281), but they are tantalizingly scrappy. As an indication

---

[1] Not only so, but it cost him a year's rates. In the Redruth Church Rate Book we find in 1798: "Capt. Richard Trevithick for Moreton's 2s. 8d.", and in 1799: "Richard Trevithick for Moretan's House, 2s. 8d."

of the extent of his practice and of his activity we may mention
that he was employed at one time or another in the following
mines: Dolcoath, Cook's Kitchen, Camborne Vean, Wheal Gons,
Wheal Bog, Trenethick Wood, Trenethick St Agnes, Wheal
Treasury, Wheal Rose, Polgine, Ding Dong, Wheal Malkin,
Rosewall, Seal Hole, Wheal Margaret, Trevenen, Wheal Hope,
Godolphin, Wheal Abraham, Wheal Prosper, Penrose, Binner
Downs, East Pell, Wheal Druid, Prince William Henry, Wheal
Clowance and Hallamannin. He was occupied with everything
required for these mines in the way of engines, pumps, boilers
and whims.

The year of Trevithick's marriage seems to coincide too with
the end of the maturing period of his life and the beginning of
the harvest of his inventions. From this time till he left for South
America was the period when he displayed the greatest brilliance,
indeed he was a veritable volcano of ideas and inventions: the
plunger pump; the water-pressure engine; the high-pressure
steam engine; the steam carriage, the steam locomotive; the steamboat; he never paused to find out whether any of his conceptions
could claim novelty—naturally not all of them could do so—
sufficient that the ideas were original with him and met the need
of the hour. The history of invention shows that originality is
rarely, if ever, associated with a mind congested with a store of
book learning, and that originality is found in the earlier years
of maturity before tradition and existing practice have had time
to damp it down. Such at any rate was the case with Trevithick.

In 1797 he began the application of the plunger pump in place
of the common bucket pump. The invention was old, for it had
been patented by Sir Samuel Morland in 1675; moreover, it had
been in use in some parts of Cornwall as early as 1786, so that
we can only claim for Trevithick that he could appreciate and
adopt a good thing, as others had done. But under his hands it
was much more than this.

The plunger pump was the central feature, or core, of the
idea out of which the Cornish engine was developed. Put
succinctly "the engine pumped the rods and the rods pumped the
water", reversing existing practice. During the steam stroke

the dead weight of the rods was lifted and the utmost advantage could be taken of expansion because a constant balance could be maintained between the acceleration and retardation of the weight of the rods and the varying pressure of the steam. An incidental advantage was that less machining is needed—turning a plunger is easier than boring a pump barrel—and a practical advantage is that a plunger is less liable to obstruction by grit than a bucket. The plunger was a cast-iron pipe plugged at its lower end and working through a stuffing-box into a pipe of somewhat larger bore, fitted with valves. It was not long before the leading Cornish mines had adopted it.

The next thing to which Trevithick turned his attention was the water-pressure engine. This again was not new, as it had been applied on the Continent and in the North and Midlands of England. Whether Trevithick knew of these applications we cannot say, but it seems not improbable that an acute mind like Trevithick's would arrive at the idea independently by mentally reversing the action of the plunger pump. His first water-pressure engine was erected at Prince William Henry (Roskear) Mine in 1798 and was followed by others, both single, with a pole, and double acting, with a piston. Fig. 2 shows one of the latter type erected by him in 1799 at Wheal Druid. In this he used an ingenious arrangement of balanced piston slide-valves. Two pistons worked, each in its own cylinder, and their spindles, passing through stuffing-boxes, were connected together by a chain passing over a sprocket-wheel. As the water pressure was constantly upon the upper sides of the pistons they were in equilibrium and therefore moved with little friction. Their necessary motion, by which the water passages were reversed to give the main piston its reciprocation, was given by a tappet and "tumbling-bob" as had been done before by Smeaton (who simply applied the known device by which the "regulator" of the Newcomen engine was actuated), but here we have an example of the ingenuity and directness of Trevithick's ideas. Some such device as an air-vessel or its equivalent was necessary to prevent shock on the closing of the valves. He did this by making his valve pistons shorter by half an inch than

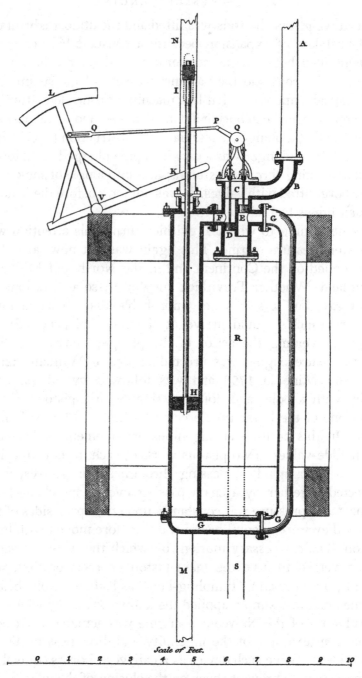

Fig. 2. Trevithick's water-pressure engine at Wheal Druid, 1799.
From *The British Encyclopedia*

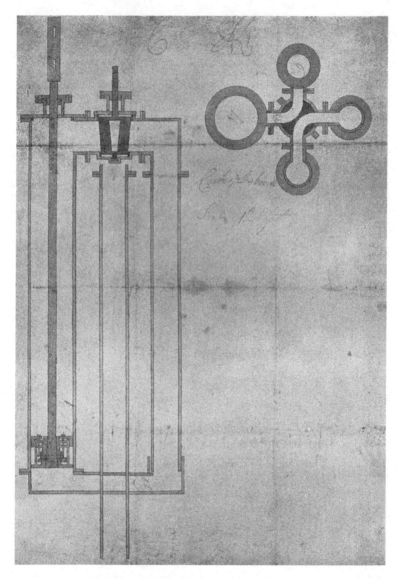

PLATE V. ORIGINAL DRAWING BY TREVITHICK OF HIS
WATER-PRESSURE ENGINE AT TRENETHICK WOOD, 1799

From the *Enys Papers*

the passages over which they moved, so that during the brief time the valves were moving, the water column was never checked enough for its inertia to come objectionably into play.

A drawing of the engine at Trenethick Wood, 1799, is in exis-

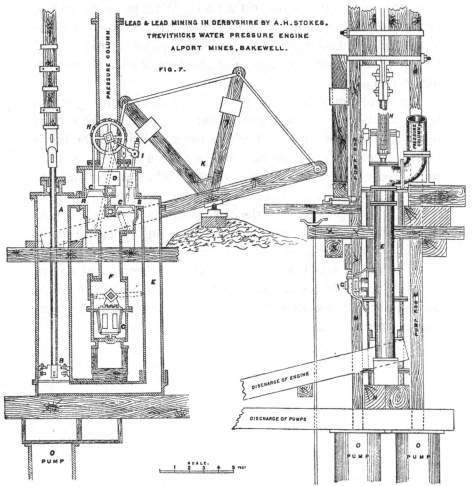

Fig. 3. Trevithick's water-pressure engine at Alport Mines, Derbyshire, 1803. From *Trans. Chesterfield and Derbyshire Mining Engineers*, 1884

tence, given to Giddy "by the Engineer Mr Richard Trevithick", and it may be his own handiwork[1] (see Plate V). The engine is of the same construction as that of Wheal Druid.

Some of his water-pressure engines were of unprecedented

[1] Preserved among the *Enys Papers*.

power. One installed at Alport Mines near Bakewell in Derbyshire in 1803 had a cylinder 25 in. diam. and a stroke of 10 ft. with a water-pressure of 65 lb. per sq. in. (see Fig. 3). Some idea of the power developed is obtained when it is noted that it was more than twice that of the 63 in. Watt pumping engine at Dolcoath. It is not too much to say that this engine was a veritable embodiment of its creator's physical and mental characteristics—strength and an exuberance of energy amounting to rashness.

The inversion of the principle of the plunger pump by which it was converted into a prime mover is typical of the directness of Trevithick's mind. Perhaps the simplicity of the high-pressure steam engine that he had conceived a few years earlier suggested to him this admirable machine which was so suitable to the requirements of mine drainage where it could be applied. At the time he was making the earliest of these water-pressure engines he was tentatively developing the high-pressure steam engine. With this development we arrive at a distinct turning-point in his life which led to his leaving the provincial sphere in which he had hitherto shone to win fresh laurels in a wider and practically international one. We pause, therefore, to commence a new chapter.

CHAPTER III

THE FIELD WIDENS

High-pressure steam—Road carriages, Camborne and London—High-pressure patent—Experimental engine at Coalbrookdale—Spread of the high-pressure engine—Explosion at Greenwich—Samuel Homfray and the South Wales locomotive—Invasion of England—Dredgers for the Thames—Thames Archway Co.—London locomotive—Sale of patent—Nautical labourer—Iron tanks—Illness—Return to Cornwall—Bankruptcy.

THE high-pressure engine, in which steam acts on a piston and is exhausted into the atmosphere after it has done its office, seems to us to-day too obvious for words; indeed there was no invention in it, nor did Trevithick claim it, for the idea was at least a century old, and Watt himself had specified it. We have rather to think of the steam engine as it then existed to realize what a fundamental change Trevithick introduced. Cylinder, condenser, air pump, ponderous beam slowly see-sawing in a massive engine house, taking months to build, fixed in its position, occupying large space, and costly—what could be said when Trevithick dispensed with the beam and air pump, acted directly on the crank-shaft, increased the speed, decreased the space needed, made it capable of being manufactured in small powers and of going anywhere and doing anything cheaply? The innovation was revolutionary and it may best be brought home to present-day readers by comparing its advent with that of the internal combustion engine.

At what date Trevithick formed the first conception of the new engine is not known, further than that it was in 1798 or possibly 1797. Giddy relates[1] Trevithick's delight at his answer to the question as to what

would be the loss of power in working an Engine by the force of Steam raised to the Pressure of several Atmospheres, but instead of condensing to let the steam escape. I, of course, answered at once that

[1] *Enys Papers. Life*, i, 63. Letter, D. Gilbert to J. S. Enys, 1839, April 29th, already quoted.

the loss of power would be one Atmosphere, diminished power by the saving of an air Pump with its Friction and in many cases with the saving of condensing water. I never saw a Man more delighted, and I believe that within a Month several "Puffers" were in actual work.

Mrs Trevithick had a vivid recollection in their early married life of her husband carrying on trials in the kitchen with models which had been made for him by her brother-in-law, William West. To these trials sundry friends were invited:[1] "A boiler something like a strong iron kettle was placed on the fire: Davies Gilbert was stoker and blew the bellows; Lady Dedunstanville was engine man and turned the cock for the admission of steam" and was charmed to see the wheels go round.

Trevithick's quick brain soon saw the possible application to locomotion: "Shortly afterwards another model was made which ran round the table, or the room. The boiler and the engine of this second model were in one piece; hot water was poured into the boiler and a red-hot iron put into an interior tube, just like the hot iron in tea-urns".

A model answering to this description is preserved in the Science Museum, South Kensington (see Plate VI). Its history can be traced back to 1810, when it was at the works of Messrs Whitehead and Co., Soho Iron Works, Manchester, who made engines for Trevithick as early as 1804. The authors believe the model to be the one and the same as that above mentioned.

A third model heated by a spirit lamp was made. "This was taken to London by a gentleman who came down for the purpose of seeing it work."[2] A surmise as to where this model went will be hazarded later. A model corresponding to this description, apparently of workmanship of the period, and possibly a copy, was picked up in Lincoln some thirty years ago and is now in the possession of Mr E. D. Lowy.

Trevithick's first application of the engine was to replace the hand windlasses and horse whims used for raising ores from the mines, a form of labour that became more and more onerous the deeper the mines went; indeed a new method of winding ore was urgently needed. It must not be supposed that Trevithick

[1] *Life*, I, 103.   [2] *Life*, I, 103.

PLATE VI. TREVITHICK'S EXPERIMENTAL MODEL ROAD
LOCOMOTIVE, 1798

Courtesy of the Science Museum, South Kensington

was the first to apply steam to winding in mines. Watt had set
up one of his sun-and-planet engines to work in 1784 at Wheal
Maid,[1] followed by several more in Cornwall and other parts of
the country. The atmospheric engine, too, was made to answer
the same purpose, albeit clumsily, by loading the beam or the
flywheel. The high-pressure engine was handier, cheaper and

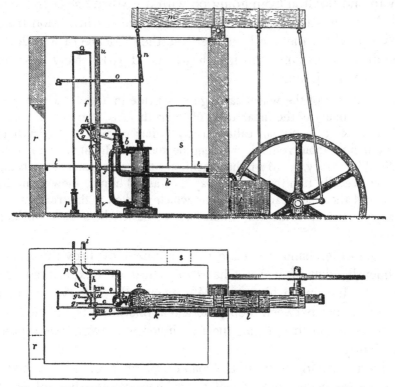

Fig. 4. Trevithick's high-pressure whim engine at Cook's Kitchen
Mine, 1800. From a drawing by Francis Trevithick

more compact than either. It received the name of "puffer"
from the noise it made, distinguishing it from its condensing pre-
decessors. The whim engine was brought into use in 1799 in
several places, and, in all, Trevithick built about thirty such
engines.[2]

[1] *James Watt and the Steam Engine*, 1927, p. 253, where drawings and
details of a number of such engines are given.
[2] *Trans. Newcomen Soc.* x, 55.

Simultaneously with the steam whim Trevithick determined, characteristically, to build a steam road carriage. The idea was in the air and his was by no means the first attempt; for example, Nicholas Joseph Cugnot made one in Paris in 1769 and, to come nearer home, William Murdock had made a model of such a carriage at Redruth in 1786, but was dissuaded by both Watt and Boulton from going on with it. Much has been made of the suggestion that Trevithick borrowed his invention from Murdock. On the one hand we have the evidence of Murdock, detailed by his son John in a letter dated 1815, May 31st, to James Watt junior:[1]

The model of the wheel carriage was made in 1792 and was then shewn to many of the inhabitants of Redruth. About two years after Trevithick & A. Vivian called at my father's house in Redruth to consult him about removing an engine then on Hallamanin mine to Wheal Treasury....My father mentions this circumstance to bring to their recollection that on that day they asked him to shew them the model of the wheel carriage engine which worked with strong steam and no vacuum. This was immediately shewn them in a working state.

On the other hand, we have the statement of Trevithick subsequently, that he had not heard of Murdock's experiment till after he had made his own. We hazard the observation that the difference between their practical applications is sufficient to justify the conclusion that the two inventions were made independently.

Before starting to build the road carriage, Trevithick realized that it was essential to see if the wheels would give sufficient adhesion on the road to propel the vehicle—a conclusion then considered more than doubtful. Giddy,[2] in the reminiscences already quoted, says:

It must have been I presume in the preceding summer [that is of 1801] that Trevithick and myself tried an Experiment on a one horse chaise, as to the hold of the Wheels on the ground for moving it up an ascent.

[1] *James Watt, loc. cit.* p. 294, where the different models made by Murdock are discussed.

[2] *Enys Papers. Life,* I, 117.

We placed the Carriage in the Middle of one of the roads leading to Camborne Church Town, and we discovered that none of the acclivities were sufficient to make the Wheels slide in any perceptible degree, as we forced the carriage forward by turning the Wheels.

Incidentally we may mention, as a sidelight on the means of transport then available, that this carriage was the only respectable passenger vehicle to be hired between Truro and Land's End and was the chaise that had been used by Watt when in Cornwall sixteen years earlier.

The construction of the road carriage was undertaken in a smith's shop belonging to John Tyack at the Weith: entries in the account books show that it was begun in November, 1800, if not earlier, but it took quite a year's experimenting, working mainly in spare time, before Trevithick had the carriage ready for the road. The momentous trial took place at Camborne on Christmas Eve, 1801. Instead of selecting the level turnpike, Trevithick set it to climb Beacon Hill, a gradient of about 1 in 20. It is facing the scene of this exploit that the Trevithick statue has been placed (see p. 262). Numbers of men climbed on the carriage and experienced for the first time the sensation of being carried along by other than animal power. Trevithick had foreseen the necessity for plenty of steam and had provided a bellows to blow the fire, but it was found insufficient and the carriage stuck on the hill. It was brought back and tried on the turnpike with, however, the same defect.

What happened further may be told in Giddy's own words, recorded in his "Pocket Book" or Diary:[1]

Dec. 27th, 1801.

I rode in the evening to Tehidy for the purpose of meeting Trevithick the next morning with his carriage.

Dec. 28th.

After waiting a long time, learnt that the Carriage had broken down.

Then follows this explanatory note, dated 1830:

A very curious sequel followed this disappointment. The Travelling engine was replaced in a building and Trevithick and Vivian and the

[1] Preserved among the *Enys Papers*.

others determined on supporting their spirits by dining at the inn. They did so and forgot to extinguish the Fire that evaporated the water and then heating the Boiler red hot, communicated fire to the wooden machinery and everything capable of burning was consumed.

D. G. 1830.

Nine years later Giddy supplied Mr J. S. Enys with another account of the fateful experiment:[1]

The Travelling Engine took its departure from Camborne Church Town for Tehidy on the 28th of Decr, 1801, where I was waiting to receive [it]. The carriage, however, broke down after travelling very well, and up an ascent, in all about three or four hundred yards. The carriage was forced under some shelter, and the Parties adjourned to the Hotel, & comforted their Hearts with a Roast Goose & proper drinks, when, forgetfull of the Engine, its Water boiled away, the Iron became red hot, and nothing that was combustible remained either of the Engine or the house.

What a delightful touch—the "roast goose" and when comforting themselves with "proper drinks" how easy to forget the boiler fire! We have never heard of more cheerful obsequies.

Another account,[2] fifty years subsequent to the event however, by Henry Vivian, nephew of Andrew Vivian, goes into greater detail:

They started to go to Tehidy House where Lord Dedunstanville lived, about two or three miles off. Captain Dick Trevithick took charge of the engine, and Captain Andrew was steering. They were going very well around the wall of Rosewarne when they came to the gully [an open water course across the road], the steering handle was jerked out of Captain Andrew's hand, and over she turned.

Rosewarne was only round the corner, so to speak, so that they had gone no distance.

Although drawings of this road carriage have appeared,[3] there

---

[1] *Enys Papers. Life*, I, 117.   [2] *Life*, I, 111.

[3] Cf. *Life*, I, 126. Francis Trevithick made them from the recollections of the old men. He followed the model in respect of the engine and boiler and on this in the authors' opinion he was right, but the framing or chassis was clearly of timber as will be inferred from Giddy's statement. It was this that so helped the fire that subsequently destroyed the vehicle; besides the fact that persons were invited to "jump on" shows that there must have been more accommodation than Francis shows.

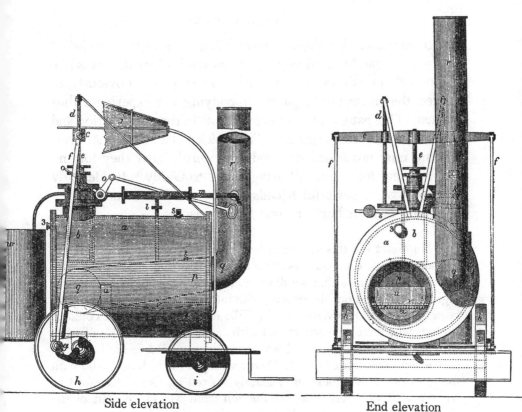

Side elevation   End elevation

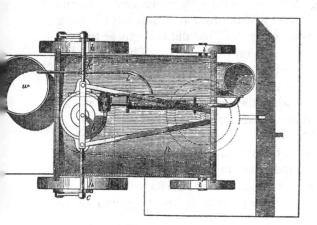

Plan

Fig. 5. Trevithick's Camborne road carriage 1802.
From a drawing by Francis Trevithick

is no authority for them. Beyond knowing that it weighed
30 cwt. and had forced draught, we can only infer that it was like
the models mentioned above. The next step was obviously to
protect the inventions by patent embodying the experience thus
gained. This patent, for "improvements in the construction and
application of steam engines", Trevithick and his cousin, Andrew
Vivian, now proceeded to London to enrol. That they lost no
time is clear from the following letter to Giddy.[1] Incidentally
it shows what powerful friends the two had got and what a
widening of their horizons was taking place.

SIR,

No doubt ere this you have been in expectation of hearing from
us, but so much time was taken up at Bristol and its Invoirns in contracting for Engines that we did not arive hear untill last wednesday
night. We waited on Mr Sandys, who informed us that the Cavit had
been renewed and Advised us to get the best information we could of
persons who were well acquainted with patents & machanick's of what
title to give the Machine and what was the intended use of it.

The next day waited on Mr Davey with your kind letter, who with
the greatest cheirfullness immediately waited on Count Rumford to
whome we had a letter of introduction from Mr D. to wait of the
Count on Friday morning. We found him a very pleasent man, and
very conversant about fire places and the Action of Steam for heating
rooms, boiling water, dressing meat, &c., but did not appear to have
studied much the Action of Steam on pistons, &c. The Count has
given us a rough draft of a fire place which he thinks is best adapted for
our Carriage, and Trevithick is now making a Compleat drawing of it.
We are to wait on the Count when the drawing is compleated, and he
has promised to give us all the assistance in his power.

Mr Davey says that a Mr Nicholson, he think[s], will be a proper
person to assist us in taking out the patent, and we are to be int[r]o-
duced to him tomorrow, and then shall immediately proceed with the
Business. We shall not specify without your Assistance, and all our
friends say that if we meet with any difficulty nothing will be so
necessary as your presence.

When we delivered Lord De Dunstanville's letter to Mr Graham,
he said he would give us every assistance in his power gratis if
wanted.

[1] *Enys Papers. Life*, p. 112. Letter dated 1, Southampton Street, Strand,
1802, Jan 16th. Holograph, A. Vivian.

Mr Pascoe Grenfill says he can find a way to the Attorney General if wanted.

It is strongly recommended to us to get a carriage made hear and to exhibit it, which we also beleive must be done.

Trevithick called on Mr Clayfield at Bristol, and is to call again on his return from Coalbrookdale to go on the mine. Both Mr Clayfields beg their most respectfull Compliments to you.

Mr Davey begins his lectures at the Royal Institution on Thursday next and has given us tickets of admittance.

We remain in good health & spirits,

Sir, Your most obliged Hble servants,
R. TREVITHICK,
A. VIVIAN.

"Mr Davey", it may be mentioned, took great interest in the steam coach and wrote:[1] "I shall hope soon to hear that the roads of England are the haunts of Captain Trevithick's Dragons. You have given them a characteristic name".

Andrew Vivian returned to Camborne and presumably Trevithick went to Coalbrookdale. Vivian wrote a chatty letter[2] full of gossip, most of which is worth quoting:

DEAR FRIEND,

I arrived here last evening safe and sound, and missing my wife, was soon informed she was at your house, where I immediately repaired. Your wife and little Nancy are very well, but Richard is not quite well, having had a complaint which many children in the neighbourhood have been afflicted with; they are a little feverish when attacked, but it has soon worn off, as I expect your little son's will also; he is much better this morning, and talked to me very cheerfully.

Mrs Trevithick is in pretty good spirits, and requested I would not say a word to you of Richard's illness, as she expected it would be soon over; but as I know you are not a woman, have given you an exact state of the facts. All my family, thank God, I found in perfect health, and all beg their kind remembrances to you, as does everyone that I have met in the village.

"How do you do?" "How is Captain Dick?" with a shake by the hand, have been all this morning employed.

---

[1] *Enys Papers*. Letter, H. Davy to D. Giddy, Royal Institution, 1802, June 12th.

[2] *Life*, p. 114. Letter dated Camborne, 1802, Feb. 23rd.

In a day or two you shall have all the particulars of mines, &c. Suffice for the time to say that Nth Downs is as good as ever. Sold 83 tons, 4th inst., chiefly halvans, at 9*l.* 18*s.* per ton, and have not, as we supposed, postponed the sampling one month, but have sampled again 124 tons.

In the Falmouth paper[1] are the following lines:

"In addition to the many attempts that have been made to construct carriages to run without horses, a method has been lately tried at Camborne, in this county, that seems to promise success. A carriage has been constructed containing a small steam-engine, the force of which was found sufficient, upon trial, to impel the carriage, containing several persons, amounting at least to a ton and a half weight, against a hill of considerable steepness, at the rate of four miles in an hour. Upon a level road, it ran at the rate of eight or nine miles an hour. We have our information from an intelligent and respectable man, who was in the carriage at the time; and who entertains a strong persuasion of the success of the project. The projectors are now in London, soliciting a patent to secure the property".

.   .   .   .   .

Pray let me hear from you on the receipt of this. When you go on with the York Water Company be sure to remember in the agreement that the new engine is not to do *more work than the old one unless paid in proportion*, otherwise they may increase their number of tenants, and keep our engine constantly at work.

With most respectful compliments to Mr and Mrs Stamp, Mary, and the little ones,

I remain, dear Sir,

Yours most sincerely,

ANDREW VIVIAN.

Mrs T. (your beloved wife) begs her love, and expects to hear from you often.

Mr Richard Trevithick, Steam Engineer,
    1, Southampton Street, Strand, London.

Apparently Trevithick now returned to Cornwall and Vivian went to town to complete the final stage in the sealing of the

---

[1] The "Falmouth paper" referred to is the *Cornwall Gazette and Falmouth Packet* then recently established. The paragraph appears in the issue of Feb. 20th, 1802.

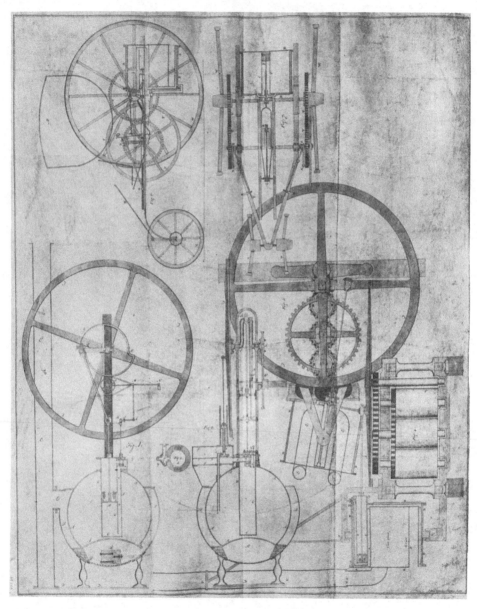

PLATE VII. DRAWING ENROLLED WITH TREVITHICK AND VIVIAN'S
PATENT FOR THE HIGH-PRESSURE ENGINE, 1802

In the Record Office

patent. The absence of Trevithick would account for the fact that the patent is signed by Vivian alone. The patent was sealed on the day following the letter and as the patent is of such far-reaching importance, being the pivot so to speak of Trevithick's career, we give it in full on another page[1] transcribed direct from the Roll in the Record Office,with a copy of the drawing attached to it. The figures, which in the original have been run into one another to keep down the size of the sheet, have been separated to render them clear; the original drawing is coloured. It will be observed that various applications of the engine are shown, viz. to road transport and to stationary purposes, and of the latter we have now to speak.

The small size of the boilers shown is particularly noticeable and it is obvious that Trevithick contemplated using forced draught generally for raising steam, yet no claim is made for this feature in the patent specification.

In August, 1802, Trevithick, who had been at Bristol, was at Coalbrookdale, where he was busy on an experimental engine in which the unprecedented pressure of 145 lb. per sq. in. was reached; the cylinder was so small that no one would credit its capabilities. The duty was ascertained by raising water by temporary pumps. He reported to Giddy on August 22nd as follows:[2]

Shod have writ you some time sence, but not haveing made sufficient tryal of the engine, have referd it untill its in my power to give you an agreeable informeation of its progress. The boiler is 4 ft Diam; the Cylinder 7 in Diam, 3 feet Stroake. The water-piston [is] 10 In, drawing and forceing 35½ feet perpendr, equal beam. I first set it off with abt 50 lb. on the Inch pressure against the steam-valve, without its Load, before the pumps was ready, and have sence workd it several times with the pumps, for the inspection of the engineers abt this nibeourhood. The steam will get up to 80 lb. or 90 lb. to the inch in abt one hour after the fire is lighted; the engine will sett off when the steam is abt 60 lb. to the inch, abt 30 Stroakes pr mt, with its load. Their being a great deal of friction on such small engine and the steam continues to rise the whole of the time its worked; it go at

[1] See Appendix, p. 269.
[2] *Enys Papers. Life,* i, 153. Letter dated Coalbrookdale, 1802, Aug. 22nd.

from 60 lb. to 145 lb. to the inch in fair working, 40 Stroakes pr mt;
it became so unmanageable as the steam encreased, that I was obligd to
stop and put a cock in the mouth of the Dischardging pipe, and leve
only a hole open of $\frac{1}{4}$ by $\frac{3}{8}$ of an inch for the steam to make its escape
in to the water. The engine will work 40 Stroakes pr mt with a pres-
sure of 145 lb. to the inch against the steam valve, and keep it con-
staintily sweming with 300 wt. of Coals every four hours. I have
now a valve makeing to put on the top of the pumps, to lode with a
Steelyard, to try how many pounds to the inch it will do to real duty
when the steam on the clack is 145 lb. to the Inch. I cannot put on
any moore pumps as they are very lofty already. The packing stands
the heat and pressure without the least ingurey whatever; enclosd
you have some for inspection that have stood the whole of its working.
As their is a cock in the dischardgeing pipe which stops the steam after
it have don its office on the piston, I judge that it is almost as fair a
tryal as if the pump load was equal to the power of the engine. Had
the steam been wire drawn between the boyler and Cylinder it wd not
have been a fair tryal but, being stoped after it is past the engine, it
tells much in its favour for haveing a greater load than it has now on.
The boyler will hold its steam a considerable time after the fire is
taken out. We worked the engine three quarters of an hour after all
the fire was taken out from under the boyler, and it is also slow in
getting up, for aftr the steam is atmosphire strong it will take half an
hour to get it to 80 or 90 lb. to the inch. Its very accomadetay to the
fire man for, fire or not, its not soon felt. The engineers abt this
place all said that it was imposiable for such small Cylinder to lift
water to the top of the pumps, and degraded the principals, tho at the
same time they spoake highely in favour of the simple and well
contrived engine. The[y] say it is a supernatureal engine for it will
work without either fire or water, and swears that all the engineers
hitherto are the bigest fools in createtion. They are constantily calling
on mee, for the[y] all say the[y] wod never beleve it unless the[y] see it,
and no persone here will take his nebiours word even if he swears to it,
for the[y] all say its an imposiabelity, and never will believe it unless
the[y] see it. After the[y] saw the water at the pump head, the[y] said
that was posiable, but the boyler wold not mantain it with steam at
that pressure five minutes; but after a short time the[y] setts off with
a solid countenance and a silent tongue.—The boyler is $1\frac{1}{2}$ In thick
and I think their will be no dainger in putting it still higher. [I] shall
not stop lodeing the engine untill the paking burn or blow under its
pressure.—[I] will write you again as soon as I have made farther
tryal. If I had 50 engines I culd sell them in a Day at any price I wod

ask for them, for the[y] are so highely pleasd with it that no other engine will pass with them.

Then in a postscript he says:

The Dale Co have begun a carrage at their own cost for the realroads and is forceing it with all expedition.

Strangely enough we do not hear any more about this engine or this locomotive. Typical of Trevithick's boldness is his statement that he does not intend to stop loading the engine until the packing burns or blows out under its pressure. His statement that if he had fifty engines ready he could sell them in a day was obviously an exaggeration, but typical of his optimism—it did look as if success was now within his grasp. It has been suggested that if, at this stage of his career, he could have been induced to have started a manufactory, and to have gathered around him some competent assistants, as Boulton and Watt had done, he might easily have attained a position of like importance. Had such an idea entered his head, and it does not seem to have done so, and been carried into practice, his restless genius, seeking ever fresh applications of his inventions, would have brought the routine of manufacture to a standstill, and the business to failure.

It will be observed that in their letter of January 16th from London, the patentees had been given the advice to build a road carriage for exhibition purposes there; perhaps they felt confident of greater success where road surfaces were less rough than in their native county. At any rate they concurred in the opinion and proceeded to build another carriage of the improved design shown in the specification. It has been suggested that a second carriage had already been made in Cornwall to replace the one set on fire at Camborne, or that that carriage had been rebuilt. The misapprehension has arisen because the engine and boiler for the second carriage were made and tried at Hayle Foundry before sending them to London. This is confirmed by Andrew Vivian's account book, which shows that in January and February, 1803,[1] a cylinder and boiler were being made at Hayle Foundry. The

[1] *Life*, i, 140.

best evidence that we have about this engine is from Trevithick himself:[1]

The coach engine did not arrive at London untill last wednesday. The coach is ready to fix the engine to. I expect we shall be ready to start in abt a fortnight. We work[ed] the engine before we sent it from home, it was perfectialy tight and [gave] double so much steam as we wanted with out blowing and the chimney was but two feet high. It worked 50 stroakes pr Mt with 5½ Inch cylinder was 30 lb. to the inch 2½ feet stroake—much more power than we shall want.

Trevithick communicated the same information to Simon Goodrich[2] of the Navy Board, Inspector General's Department. Goodrich made a memorandum of the interview, but unfortunately did not date it. He gives additional information and best of all accompanies it by a sketch. Goodrich's memorandum is as follows:

Mr. T. has prepared an Engine for drawing a coach, diam. of Cylinder 5½ stroke 2 ft. 6 in. 50 strokes per min't work on each side piston 20 lbs. pr inch in the boiler 30 lb —about 110,000 lb. 1 ft. high in a minute. The whole engine fire and all is 6 cwt. The whole is contained within a cylinder boiler—2.9 long 2.6 diam.

Hyde Clarke in 1839 wrote: "The engine he used was about the size of an orchestra drum", not an inapt description. The coachwork was supplied by William Felton of Leather Lane, a street running between Holborn and Clerkenwell Road. We know this from Andrew Vivian's account book:

July 1803. To Felton for building the coach   83l. 5s. 0d.

Beyond the fact that Felton had been in business in Oxford Street in 1797, was in Leather Lane in 1802 and was not there in 1805, we know nothing about him.[3] Obviously 83l. was not the whole cost of the coach and can only have been a payment in

[1] *Enys Papers. Life*, p. 158. Letter dated Bristol, 1803, May 2nd.

[2] For information about Goodrich (1773–1847) and a full discussion on this engine, see *Trans. Newcomen Soc.* 1, 47.

[3] Except that he was the inventor of a warming pan for coaches: *Home Office Warrants for Inventions*, 11th Jan. 1802, MS., but it did not get to the Great Seal.

advance or on account. The total cost was 207*l*., as we deduce from Trevithick's own accounts.

The engine and boiler were sent by sea from Falmouth, as shown in Vivian's accounts:

Aug. 1803. To paid Messrs. Foxes shippers Falmouth for carriage of the engine to London 20*l*. 14*s*. 11*d*.

Clearly then the assembling was done at Leather Lane. Captain Henry Vivian stated[1] in 1845 that his father Andrew Vivian

worked the engine when it ran from Leather Lane from the shop of Mr Felton (who built the carriage, and he and his sons were with the engine all the first day it ran) through Liquorpond Street into Gray's Inn Road by Lord's Cricket Ground to Paddington and Islington and back to Leather Lane.

We have the reminiscences, about sixty-five years after the event, of Captain John Vivian of H.M. packet service. He was a young man of nineteen at the time and his recollection is likely to have been vivid, particularly as he drove the carriage. He[2]

thinks the engine had one cylinder, and three wheels; the two driving wheels behind were about 8 feet in diameter. The boiler and engine were fixed just between those wheels. The steering wheel was smaller, and placed in front. There were some gear-wheels to connect the engine with the driving wheels. The carriage for the passengers would hold eight or ten persons, and was placed between the wheels, over the engine, on springs. One or two trips were made in Tottenham Court Road, and in Euston Square. One day they started about four o'clock in the morning, and went along Tottenham Court Road, and the New Road, or City Road: there was a canal by the side of the road at one place, for he was thinking how deep it was if they should run into it. They kept going on for four or five miles, and sometimes at the rate of eight or nine miles an hour. I was steering, and Captain Trevithick and some one else were attending to the engine. Captain Dick came alongside of me and said, "She is going all right". "Yes," I said, "I think we had better go on to Cornwall." She was going along five or six miles an hour, and Captain Dick called out, "Put the helm down, John!" and before I could tell what was up, Captain Dick's foot was upon the steering-wheel handle, and we were tearing down

---

[1] *Life*, I, 140. [2] *Life*, I, 142.

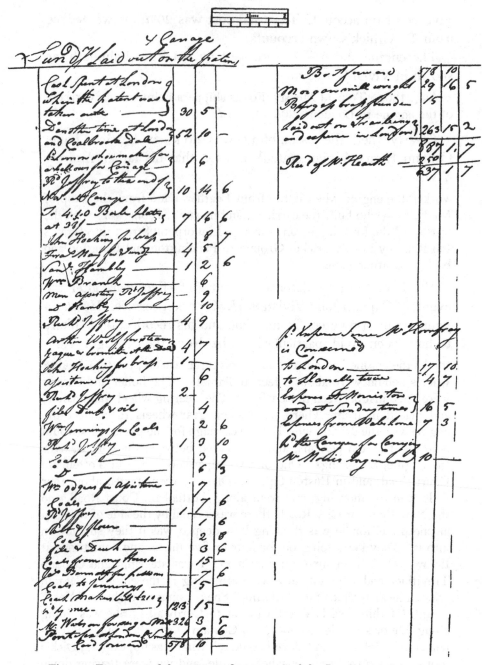

Fig. 6. Expenses of the patent of 1802 and of the London road carriage.
From Trevithick's Account Book

six or seven yards of railing from a garden wall. A person put his head from a window, and called out, "What the devil are you doing there! What the devil is that thing!"

They got her back to the coach factory. A great cause of difficulty was the fire-bars shaking loose, and letting the fire fall through into the ash-pan.

The waste steam was turned into the chimney, and puffed out with the smoke at each stroke of the engine. When the steam was up, she went capitally well, but when the fire-bars dropped, and the fire got out of order, she did not go well.

What is sufficiently remarkable is the fact that no notice of a sight which was, after all, an extraordinary and novel one should have appeared in the newspaper press of the day or elsewhere. The experiments were promising, indeed we can say successful, but the parties had spent a great deal of money on them, and no return seemed to be in sight, while the demand for the stationary engine was increasing rapidly. Hence they relinquished further trials and thus ended Trevithick's connection with steam carriages on common roads.

Before the trials were over Trevithick had, in April, 1803, an engine in London boring brass cannon, and another pumping water out of the foundations of a corn mill under construction on the Thames between Greenwich and Woolwich. He had another in Derbyshire,[1] besides the one at Coalbrookdale and others in prospect.

But now a most unfortunate accident happened—the explosion of one of his boilers at Greenwich on September 8th, 1803. This accident had a material influence on the introduction of the new engine, for it was bruited abroad[2] to his detriment far and wide and much misrepresentation was made of the accident although it was due entirely to carelessness. Trevithick went as soon as he could to inspect it and we can best give the report in his own words in a letter to Giddy:[3]

I found it burst in every direction. The bottom stood whole on its

---

[1] *Life*, I, 158. Letter dated May 2nd.
[2] E.g. in Tilloch's *Phil. Mag.* XVI, 1803, p. 372.
[3] *Enys Papers. Life*, II, 124. Letter dated Penydarren, 1803, Oct. 1st.

seating; it parted at the level of the chim[e]. The boiler was cast iron of abt one Inch thick, but some parts were near 1½ Inch; it was a round boiler, 6 feet Diam; the cylinder was 8 inch Diamr working double; the bucket [was] 18 Inches [diameter], 21 feet column, working single, from which you can judge the pressure required to work this engine. The pressure, it appears, when the engine burst, must have been very great, for their is one pice of the boiler, its abt one Inch thick and abt 500 lb, thrown upwards of 125 yards; and from the hole which it cutt in the ground on its fall, it must have been nearly perpendiculear and from a very greath height, for the hole it cutt was from 12 to 18 In Deep. Some of the bricks was thrown two Hundred Yards, and not two bricks [were] left fast to each other, either in the stack or round the boiler.... It appears the boy that had the care of the engine was gon to catch eales in the foundeation of the building, and left the care of it to one of the Labourers; this labourer saw the engine working much faster than usual, stop'd it without takeing off a spanner which fastned down the steam lever, and a short time after being Idle it burst. It killed 3 on the spot and one other is sence dead of his wounds. The boy returned at the instant and was then going to take off the trigg from the valve. He was hurt, but is now on recovery; he had left the engine about an hour....

...I beleive that Mr B. & Watt is abt to do mee every engurey in their power for the[y] have don their outemost to repoart the exploseion both in the newspapers and private letters very different to what it really is....

The boiler was, as we have seen, of cast-iron 6 ft. diam.; no doubt it resembled closely that described and illustrated in Figs. 1 and 2 of the patent specification (Plate VII). Trevithick asked Giddy to calculate what pressure was required to burst the boiler and disperse the fragments as described, but it does not appear that he attempted to do so.

In a marginal note to this letter, Trevithick says:

I shall put two steam valves and a steam guage in future, so that the quicksilver shall blow out in case the two valves sho'd stick and all the steam be discharged through the guage. A small hole will discharge a great quantity of steam at that pressure.

This intention was carried out as we learn from Farey, who says that[1] as a result of the accident "Trevithick took the pre-

[1] *Steam Engine*, vol. II (unpublished), in Patent Office Library.

caution in subsequent engines to have two separate safety valves and even proposed to have one locked up". Farey also mentions the lead rivet for the part of the boiler or flue exposed to the direct action of the fire, i.e. the precursor of the boiler safety plug, and says:

> By these precautions Mr T. regained so much confidence, as to obtain some orders for high pressure engines in London; but not as many as he would have received if the explosion at Woolwich had not deterred many persons from adopting his engine.

The boiler—the all-important part of the plant—is shown in the patent specification in two forms, one globular externally fired, and the other cylindrical with a "breeches" or return flue internally fired. By this time the shell was of cast-iron, and the internal flue of wrought-iron, while the cylinder was sunk in the boiler so that the arrangement was most compact.

In the letter to Giddy from Bristol, of May 2nd, 1803, mentioned above, Trevithick says that he has "sold to a Gentleman of this place one Quarter part of the patent for ten thousand pounds but this must remain a secret"; by "of this place" Trevithick probably meant someone whom he had met there. Six months later we find Trevithick, in the letter already quoted, saying:[1]

> Mr Homfray of this place have taken mee by the hand...their is ordered above 700 Horse powers at £12 12s. for each Horse power for the patent right, and the person that order them make them themselves without any expence to me whatever. If I can be left quiet a short time I shall do well.

There cannot be much doubt that the "gentleman of this place" and "Mr Homfray" were one and the same person. Samuel Homfray (1761–1822), of Penydaren Ironworks, ironmaster—one of the three sons of Francis Homfray, pioneer of the South Wales coal trade—was an able business man, brimful of energy and fond of cards and horse-racing. He was admitted to a half share in the patent and was soon pushing the engine vigorously himself.

---

[1] *Enys Papers. Life*, II, 125. Letter to Giddy, dated 1803, Oct. 1st.

Trevithick was extraordinarily active. Vivian's[1] account book shows that orders were placed with engine-makers up and down the country, and that patent premiums were paid on a large number of engines for winding, puddle rolls, forge, blowing blast furnaces, pumping and shaft sinking, in such different places as

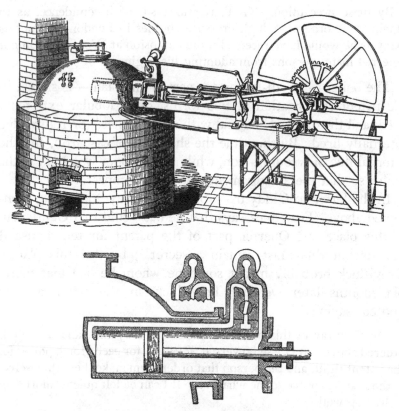

Fig. 7. Trevithick's high-pressure engine and boiler, 1803.
From Farey's *Steam Engine*, vol. II (unpublished)

Cornwall, Coalbrookdale, Tredegar, Bridgnorth, Stourbridge, Manchester, London—indeed there was hardly any purpose to which Trevithick would not apply the engine, or place where he would not go.

His restless genius had already proposed the application to locomotion on common roads; now the tramways in use in the

[1] *Life*, I, 227.

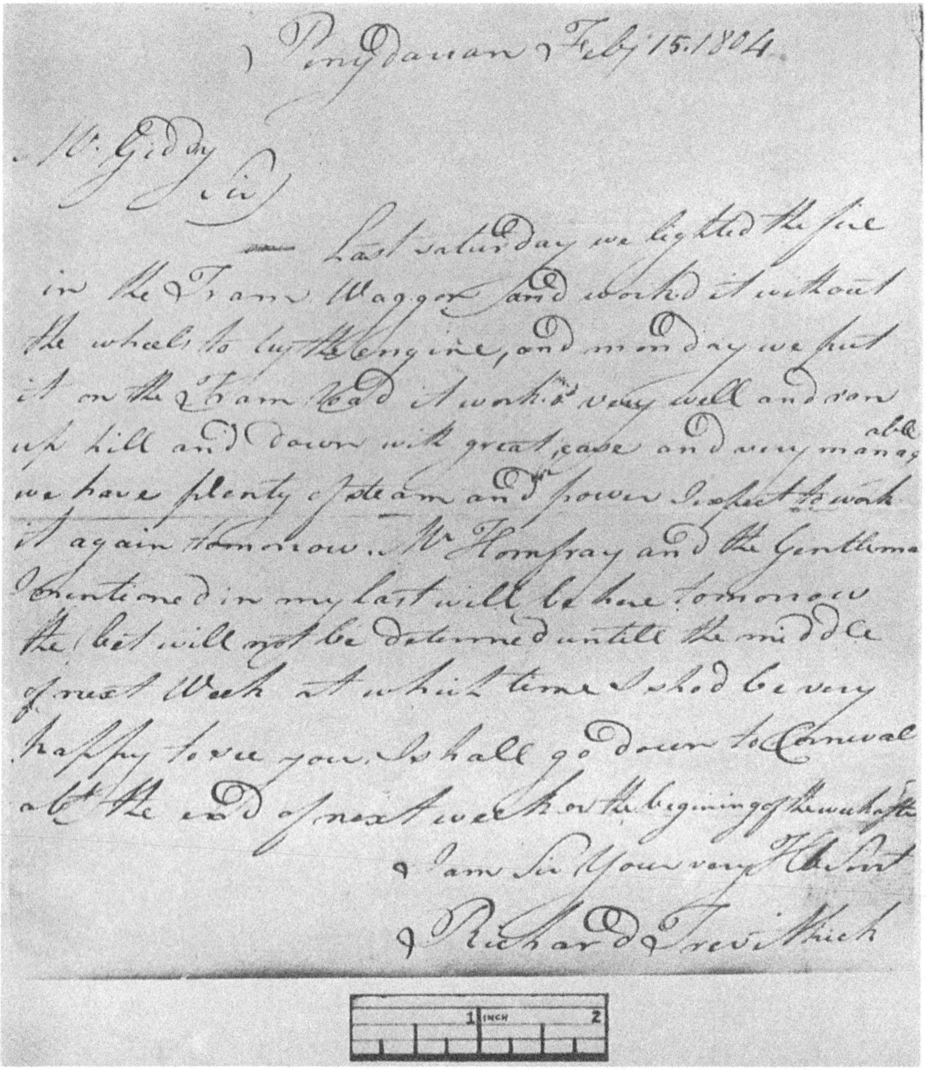

Pengdarran Feby 15. 1804

Mr Giddy
Sir

Last saturday we lighted the fire in the Tram Waggon and worked it without the wheels to try the engine, and monday we put it on the Tram Road It work'd very well and ran up hill and down with great ease and very many able we have plenty of steam and power I expect to work it again tomorrow. Mr Homfray and the Gentlemen I mentioned in my last will be here tomorrow the bet will not be determined untill the middle of next Week at which time I shod be very happy to see you. I shall go down to Cornwal abt the end of next week or the begining of the week after

& I am Sir Your very Hble Svt

Richard Trevithick

PLATE VIII. HOLOGRAPH LETTER FROM TREVITHICK
TO GIDDY, 1804

From the *Enys Papers*

district at once suggested to his mind the application of his engine to drawing loads on a confined track. Homfray, at whose works Trevithick had already built stationary engines, offered him facilities for carrying out the experiment in which he himself became much interested. In discussion among the neighbouring ironmasters, Homfray, who was of a sporting turn of mind, made a bet with Anthony Hill, proprietor of the Plymouth Ironworks, that he would haul ten tons of iron on the tramway by means of a steam engine from Penydaren to the basin at Abercynon on the Glamorganshire Canal, a distance of $9\frac{3}{4}$ miles. Hill bet Homfray even money that it could not be done. Of the difficulties that were encountered and how Trevithick surmounted them, we know nothing. It was not till the locomotive was ready that we have a definite account of it, and this it will be well to give in Trevithick's own words. He had reached Penydaren and wrote from there to Giddy (see Plate VIII).[1]

Last saturday (i.e. Feb. 11th) we lighted the fire in the Tram Waggon and work'd it without the wheels to try the engine; on Monday we put it on the Tram Road. It work'd very well, and ran up hill and down with great ease, and [was] very managable. We have plenty of steam and power. I expect to work it again tomorrow. Mr Homfray and the Gentleman I mentioned in my last, will be home tomorrow. The bet will not be determed untill the middle of next Week, at which time I shod be very happy to see you.

Five days later Trevithick writes:[2]

The Tram Waggon have been at work several times. It works exceeding well, and is much more managable than horses. We have not try'd to draw but ten tons at a time yet, but I dought not but we cou'd draw forty tons at a time very well for 10 tons stands no chance at all with it. We have not been but two miles on the road and back again, and shall not go farther untill Mr Homfray comes home. He is to dine [at] home to-day, and the engine will go down to meet him. The engineer from Government is with him. The engine, with water encluded, is ab't 5 tons. It runs up the Tram road of 2 Inch in a Yard 40 Stroakes pr min't with the empty waggons. The engine moves forth 9 feet [at] every stroake. The publick is much taken up with it.

[1] Enys Papers. Life, I, 159. Letter dated Penydarran, 1804, Feb. 15th.
[2] Enys Papers. Life, I, 160. Letter dated 1804, Feb. 20th.

The bet of 500 Hund^d Guineas will be desided abt the end of this Week, and your pressence wod give mee moore satisfactn than you can consive, and I dought not but you will be fully satisfyde for the toil of the journey by a sight of the engine. The steam thats disscharged from the engine is turned up the chimney abt 3 feet above the fire, and when the engine is working 40 St pr mt, 4½ ft Stroake, Cylinder 8¼ In Diam, not the smallest particle of steam appears out of the top of the chimny, tho' the Chimny is but 8 feet above where the steam is deliverd into it, neither is any steam at a distance nor the smallest particle of water to be found. I think its made a fix'd air from the heat of the Chimny. The fire burns much better when the steam goes up the Chimney than what it do when the engine is Idle. I intend to make a smaller engine for the road, as this has much moore power than is wanted here. This engine is to work a hammer. Their will be a great number of experiments tryd by this engineer from London respecting these engines, as that is his sole buisness here and that is my reason for so much wishing your here. He intends to try the strength of the boiler by a forceing pump, and have sent down orders to get long steam guages and forceing pumps ready for that purpose against he arrive. We shall continue our journey on the road today with the engine, untill we meet Mr Homfray and the London engineer, and intend to take out the horses out of the coach and fasten it to the engine and draw them home. The other end of the tram road is 9¾ miles from here. The coach axels are the same length as the engine axels, so that the coach will run very easely on the tram road. There have been several experiments made by Mr Homfray and this engineere latly in London on these engines. I am very much obligd to you for your offer in assisting to make out a publication of the duty and advantages of these engines. As soon as I can get proper specimens at work, and you as an eye-witness to the performance, [I] shall value your kind offer and assistance far beyond any other to be got, as you have been consulted, and [have] assisted mee from the beginning.

Two days later Trevithick announced that the trial had taken place the preceding day, Tuesday, February 21st, 1804, a date for ever memorable in the history of the locomotive:[1]

Yesterday we proceeded on our journey with the engine; we carry'd ten tons of Iron, five waggons, and 70 Men riding on them the whole of the journey. Its above 9 miles which we perform'd in 4 hours & 5 Mints, but we had to cut down som trees and remove some Large

[1] *Enys Papers*. Letter to Giddy, dated Penydarran, 1804, Feb. 22nd.

rocks out of road. The engine, while working, went nearly 5 miles pr hour; there was no water put into the boiler from the time we started untill we arriv'd at our journey's end. The coal consumed was 2 Hund^d. On our return home, abt 4 miles from the shipping place of the Iron, one of the small bolts that fastened the axel to the boiler broak, and let all the water out of the boiler, which prevented the engine returning untill this evening. The Gentleman that bet five Hund^d Guineas against it, rid the whole of the journey with us and is satisfyde that he have lost the bet. We shall continue to work on the road, and shall take forty tons the next journey. The publick untill now call'd mee a schemeing fellow but now their tone is much alter'd.

Not satisfied with having won the bet for Homfray, although as we shall see later Mr Hill at first refused to pay because of some quibble, Trevithick now aimed at making the engine a *multum in parvo*, i.e. to lift water, to work a hammer, to wind coal and then to travel on rails, all for the benefit of an engineer from the Admiralty, who was to come down with Mr Homfray to inspect. We know, from evidence to be adduced later, that this engineer was none other than Simon Goodrich,[1] mechanist to the Navy Board under Sir Samuel Bentham.

Trevithick announced the completion of these preparations in a letter to Giddy, as follows:[2]

We have tryd the Carrage with Twenty five tons of Iron and find that we were more than a match for that weight. We are now pre-pareing to get the matts [i.e. materials] ready for the experements for the London engineers who is to be here on sunday next. We have 28 feet of 18 Inch pumps fix'd for the engine to lift water as those engeneeres particulearly requested that they might have a given weight lifted, so as to be able to calculate the real duty don by a bushell of coal. Its the waggon engine that is to lift this water, then go from the pump itself and work a hammer, and then to wind coals and lastly to go the journey on the road with Iron. We shall have all the work ready for them by the end of the week. They intend staying here abt 7 or 8 days, and as the report that they will make on their return will be the standing or the condeming those engines, its my reason for so ancxiously requesting your presence. For as the[y] intend to make tryal of the Duty performed by the Coals consumed,

[1] See *Trans. Newcomen Soc.* III, 1, for an account of him.
[2] *Enys Papers. Life*, I, 166. Letter dated 1804, March 4th.

they will state it as against the duty performed by Bolton's great engines which did upwards of 25,000,000, when their 20 Inch Cylinders, after being put in the best order posiable, did not exceed 10,000,000. Therefore as you was consulted on all those tryals of Bolton's engines, your presence wod have great weight with those Gents, otherwise I shall not have fair play. Let mee meet them on fair grounds and I will soon convince them of the superiority of the pressure-of-steam engines.

The steam is deleverd in to the Chimney above the damper, and when the damper is shut the steam then makes its appearance at the top of the chimney, but when open none can be seen. It makes the draft much stronger by going up the Chimney; there is no flame that appairs. The Coals here have but very little bitteman in it, therefore but very little smoke gets from it. We never tryd a torch at the top of the chimney.

Perhaps there may never be such an opportunity when your assistance in those experiments will be of so great a benefit to mee as at this time, therefore I hope you will forgive me for again Requesting your attendance on a business that may be of such consiquence to me.

The statement in this letter about delivering the exhaust steam into the chimney and the observation that "it makes the draft much stronger" is the first unmistakable reference to the origin of that important adjunct to steam engineering in general and to the steam locomotive in particular. Trevithick was undoubtedly its originator and might have secured a master patent for it, but the device was in advance of the times.

Unfortunately Mr Homfray had an accident which put off the proposed inspection, hence Trevithick wrote to Giddy to tell him that the matter was not now urgent. The letter is as follows:[1]

I am sorry to inform you that the experements that was to be exhibited before the London gents is putt off on account of an axident which happ'ned to Mr Homfray on tuesday last. His horse ran away with the gig and threw him out & hurt his face, sprain'd his ankle, and disslocated his arm at the elbow. The experements will go on as usal; and every thing is nearley ready for it but if you com at this time you will not be favored with either Mr Homfray's Compainy or the London Gentlemen. Perhaps it may be one Month before these gentlemen will come down, and they are the persons I wish you to see. But at all events I hope you will be so good as to call this way before

[1] *Enys Papers. Life*, I, 168. Letter dated Penydaren, 1804, March 9th.

[you] go to Cornwall, as all the apperatus is nearley ready. We can go through the whole of the experements at any time....I find myself much dissappointed on account of the axident for I was very desireous to make the engine go through the different works that its effect might be publish'd as earley as posiable....I shod be very happy to see any Gentleman that you cu'ld recomend this way for informeation, as the more publick its made the sooner the engines will circulate. We have not made any experements since I writ you last. I rec'd a letter from home this morning saying that they had seen the steam-carriage in the newspaper, but did not believe it to be truth....I cannot see any release for me from this place soon, and intend to go down almost immedtly to Cornwall and bring up my family to spend the next summer here.

However, before he got this letter Giddy had set off for Penydaren. His account, written thirty-five years later, is as follows:[1]

Being in Oxford with my Father and Sisters in the Winter & Spring of 1804, I was earnestly entreated by Trevithick and by Mr Samuel Homfry, a great Ironmaster of Merthar Tidwell to come there & assist them in some Experiment. I according[ly] left Oxford & reach[ed] Merther Tidwell on the 12th of March, & on the 24th the Engine which Trevithick had constructed for going on the Rail Way, travelled from Pen-y-darren, Mr Homfray's, to works called Plymouth & back again; but all the weight was accumulated on the same four Wheels, with the Engine, for none of us once imagined, if the weight were divided, that the Wheels of the Engine Carriage could possibly hold. In consequence of this great pressure, a large number of rails broke, & on the whole the Experiment was considered as a failure.

After Giddy's departure the locomotive made at least two more trips, as shown in a letter from Homfray to Giddy on July 10th and from Trevithick to the same on July 5th.[2]

The Tram engine have carry'd two Loads of 10 tons of iron to the shipping place since you left this place, but Mr Hill says that he will not pay the bet, becase there were some of the tram plates that was in the tunnel removed so as to get the road in the middle of the arch.

[1] *Enys Papers. Life*, I, 117. Letter dated Eastbourne, 1839, April 29th.
[2] *Enys Papers. Life*, I, 170, 171. Letter dated Stourbridge, 1804, July 5th.

The first objection he started was that one man shod go with the
engine without any assistance, which I preformd myself without help;
and now his objection is that the road is not in the same place as when
the bet was made. I expect Mr Homfray will be forced to take steps
that will force him to pay. As soon as I return from here their will
be another tryal, and some person will be calld to testify its effects,
and then I expect their will be a law suite immidly. The Travleing
engine is now working a hammer.

Apparently the engine was being kept in working order in
anticipation of the visit of the "London engineers". These
gentlemen were Goodrich, already mentioned, and Lieut. Cunningham of Woolwich, representing the Navy Board, for from
Goodrich's *Journal* under March 9th, 1804, we learn that "Mr
Homfray, Iron Founder, from Pendarron Works in Wales" had
invited him to witness experiments with Trevithick's engines.
Two letters from Homfray to Goodrich are preserved agreeing
substantially with Trevithick's to Giddy in dealing with these
trials and the reason for their postponement. There is no record
of a visit and the trials must have been put off for good; hence
we are deprived of a contemporary account which would have
been of the utmost value, for we do not really know what the
engine was like.

Unfortunately we have no authentic drawing of the engine. A
sketch purporting to represent it was brought to light in 1858
and is now preserved in the Science Museum, South Kensington
(see Plate IX). This sketch bears evidence of having been
copied from a previous one. Measured to any ordinary scale,
it does not agree with the known dimensions of the Penydaren
or of the Coalbrookdale locomotive. Details are obviously not
to scale. Nevertheless the authors believe that the sketch can be
taken as being substantially a representation of the Penydaren
locomotive.[1]

Trevithick's remark that an account had appeared of "the
steam-carriage in the newspapers" must refer to the report that
appeared in the *Cambrian* newspaper of February 24th, 1804;

[1] The evidence is very fully sifted in a paper by Mr W. W. Mason,
Trans. Newcomen. Soc. XIII.

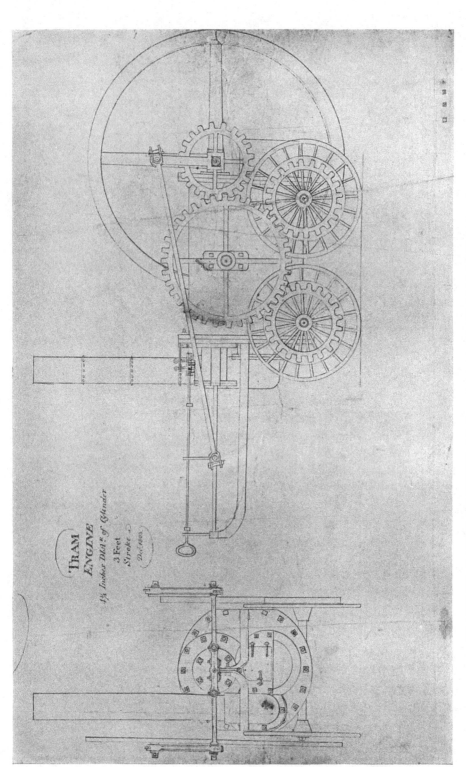

TRAM ENGINE

4¾ Inches Diam.r of Cylinder

3 Feet Stroke

Oct.r 1803

PLATE IX. EARLY DRAWING OF TREVITHICK'S TRAM LOCOMOTIVE, 1803

Courtesy of the Science Museum, South Kensington

unfortunately it gives no technical information, but is worth quoting:

Extract of a letter from Merthyr Tidvil, Feb. 22nd, 1804.

Yesterday the long-expected trial of Mr Trevithick's new-invented steam-engine, for which he has obtained his Majesty's letters patent, to draw and work carriages of all descriptions on various kinds of roads, as well as for a number of other purposes, to which its power may be usefully applied, took place near this town, and was found to perform, to admiration, all that was expected from it by its warmest advocates. In the present instance, the novel application of Steam, by means of this truly valuable machine, was made use of to convey along the Tram-road ten tons long weight of bar-iron from Penydarren Iron Works, to the place where it joins the Glamorganshire Canal, upwards of nine miles distance and it is necessary to observe that the weight of the load was soon increased from ten to fifteen tons, by about seventy persons riding on the trams, who, drawn thither (as well as many hundreds of others) by invincible curiosity, were eager to ride at the expence of this first display of the patentee's abilities in this country. To those who are not acquainted with the exact principle of this new engine, it may not be improper to observe, that it differs from all others yet brought before the public, by disclaiming the use of condensing water, and discharges its steam into the open air, or applies it to the heating of fluids, as conveniency may require. The expence of making engines on this principle does not exceed one half of any on the most improved plan made use of before this appeared; it takes much less coal to work it, and it is only necessary to supply a small quantity of water for the purpose of creating the steam which is a most essential matter. It performed the journey without feeding the boiler or using any water, and will travel with ease at the rate of five miles an hour. It is not doubted but that the number of horses in the kingdom will be very considerably reduced, and the machine in the hands of the present proprietors, will be made use of in a thousand instances never yet thought of for an engine.

Trevithick's own comment when writing in 1812 to Sir John Sinclair of the Board of Agriculture, a great patron of improvements in the arts,[1] is "I thought this experiment would show to the public quite enough to recommend its general use: but though promising to be of so much consequence, has so far remained

[1] *Life*, ii, 42, 43.

buried which discourages me from again trying at my own expense", and further on he says: "the first and only self-moving machine that ever was made to travel on a road with 25 tons at four miles per hour and completely manageable by only one man, I think ought not to be dropped without further experiments".

In the authors' opinion Trevithick was quite right in not allowing himself to be deflected from the main current of his engine development in which he was now plunged, but it can be said safely that had he had the time and the incentive to do so, he could have anticipated the development of the locomotive by fifteen or twenty years. The fact remains that Trevithick's was the pregnant idea that in other hands led to that development; as confirmation of this statement we must mention the locomotive engine built from his instructions at Newcastle and tried in 1805, for from this in lineal descent came the Wylam and the early Stephenson locomotives. Drawings of this locomotive have been preserved and are now in the Science Museum, South Kensington. The locomotive, it will be seen (Plate X), is almost identical in design with the Penydaren one. It appears that the engine was made at the foundry of John Whinfield, Pipewellgate, Gateshead, with perhaps the aid of some parts from South Wales, by John Steel, a millwright employed by Trevithick. It was intended for Mr Blackett of Wylam Colliery and was to have run on wooden edge rails. It was tried at Gateshead but was never put into service, possibly it broke the rails owing to its weight, which was 4·5 tons. The cylinder was 9 in. diam. and stroke 3 ft.

We have devoted perhaps a disproportionate amount of space to Trevithick's locomotive experiments and have not sufficiently stressed the everyday work he was doing in introducing his engine. The reason why he went into South Wales and met Homfray was because the latter wanted engines for his ironworks and collieries. Such engines were put down at Penydaren, Tredegar and other places and remained in use for half a century.[1] As evidence of what Trevithick was doing, we quote an extract from a letter which is important because it shows that he had

[1] *Life*, i, 222.

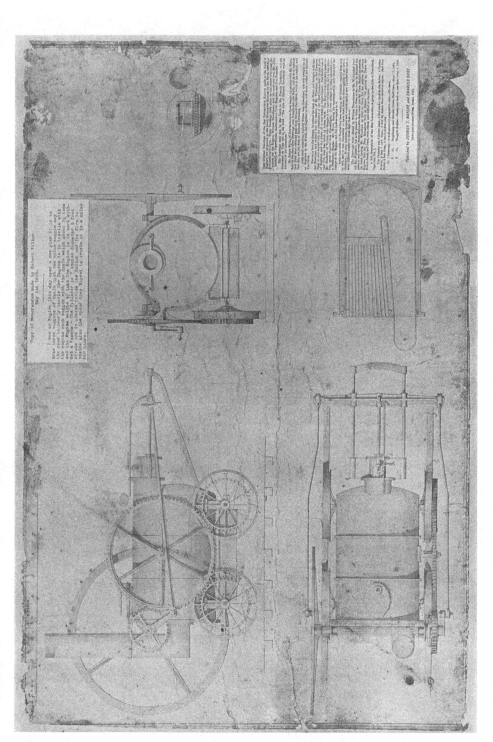

PLATE X. ORIGINAL DRAWING OF TREVITHICK'S NEWCASTLE LOCOMOTIVE, 1805
Courtesy of the Science Museum, South Kensington

already grappled with the problem of making his engine work expansively:[1]

The great engine at Penydarran goes on exceeding well. The engine will rolle 150 Tons of Iron a Week with 18 Tons of coals and the two engines of Boltons at Dolas burns 40 Tons to rolle 160 Tons; the[y] are a 24 Inch and a 27 In. Double. That at Penydarran is

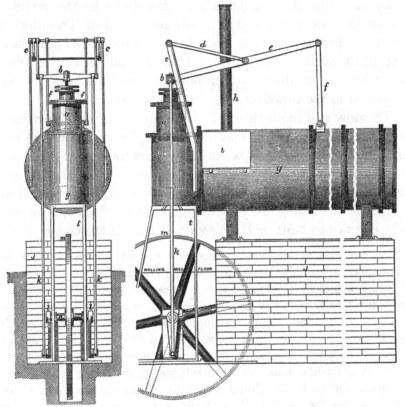

Fig. 8. Trevithick's engine for driving puddle rolls at Tredegar, 1804 to 1856. From a drawing by Francis Trevithick

18½ Inch 6 ft. Stroke, works abt. 18 St. pr minute; it requires the steam abt. 45 pounds to the Inch above the atmosphire. I workd it expancive first and found that when working the hammer which was a moore regulear load then rolling with steam high enough to work 12 St. pr. Mt. with the cock open all the stroak. Then I shutt it of at half the stroake which reduced the number of stroakes to 10½ pr. Mt.

[1] *Enys Papers*. Letter to Giddy, dated Sturbridge, 1804, July 5th.

and the steam and lode the same in both but I did not continue to work
it expancive becase the work in rolling is very uneven and the workmen wod stop the engine when working expancive, but when the
Cylinder was full of steam the Rollers culd not stop it, and as coals is
not an object here, Mr Homfray wishd the engine might be workd to
its full power.

By this time the boiler had settled down to the internal
return flue construction which was so typical of Trevithick's
practice. In the particular case shown by the illustration (see
Fig. 9) the shell is of cast-iron and the internal flue of wrought-
iron; later he made the whole boiler of wrought-iron, as ex-
emplified in the thrashing engine (see p. 131).

To show still more clearly how deeply immersed Trevithick
was at this juncture in the task of supplying his stationary engine,
and how impracticable it would have been for him to have devoted time to anything else—for he never seems to have entrusted or delegated his work to others—we feel constrained to
quote the following very long letter which describes nearly
fifty engines as built or building up and down the country:[1]

I left Wales ab^t 8 Weeks since and put an engine to work in
Worcester, a ten horse power for Driving two pair of grist stones and
carrying a leather Dressing machine, one other in staffordshire for
Winding Coals, both works exceeding well. From Coalbrooke Dale
I went to Liverpool where a founder had made two of them which
work^d exceeding well one other was nearley finishd and three others
begun. There was some spanish merchants there and saw one of them
at Work and said that as soon as he returned to spain that he wod send
an order for 12 of 12 Horse power for south America. In south
America and the spanish West India water is very scarce and in
several places scarsley water for the inhabetance to drink, therefore
there is no water for any engine whatever, but by making enquirey I
found that ten Mules wod Rolle as much cane in an hour as wod
produce 250 Gall^{ns} of the Cane juce and that juce will produce one
pound of sugar from one Gallon of juce which the[y] boile untill the
Water is all avaperated and the sugar produced. I told them that the
engine boiler might be feed with this juce and have a cock in the

---

[1] *Enys Papers.* Letter, Trevithick to Giddy, Coalbrookdale, 1804, Sept.
23rd.

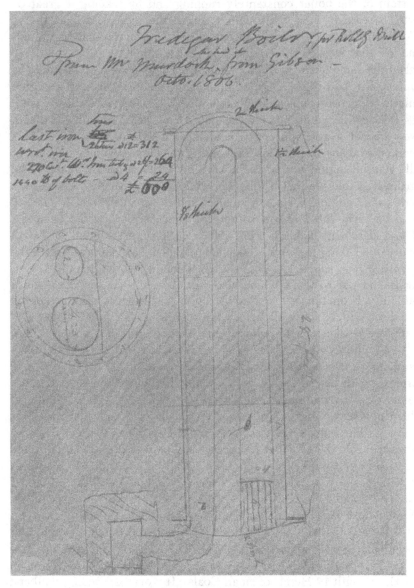

Fig. 9. Trevithick's boiler for the rolling mill at Tredegar, 1806.
From the Boulton and Watt Collection, Birmingham

bottom of the boiler constantely running and by takeing a great or small stream from it the[y] might make the juce as rich as the[y] liked. In this process, as the juce wod be so far on towards sugar, and the fire that worked the engine wod cost nothing becase it wod have taken the same quantity of fuel under the sugar pans to avaporate the water as it wod in the engine boiler, and the steam from the engine might be turned round the outeside of the furnace to destill the rum. As the destillerys do not require but a slow heat, I think the steam wod answer a good purpose round the out side of the pan. If this method wod answer, the cost of working the engine wod be nothing at all even the engine wod be there working with out fire or Water. The spainyards told me that if this plan wod answer, that the[y] wod take a thousand engines for south america and the spanish west Inidias. I wod be verry much obligd to you for your opinion on the buisness. These merchants makes a trade of buying up sugar mill and pans with every other thing the[y] want from england, and exchainges it with the spanyards for sugar. At Manchester I found two engines had been made and put to work which also workd very well and three more building. From thence I went to Derby shire: the great pressure engine I expect will be at work before the middle of October. There is a foundrey at Chesterfield building one steam engine as a sample; there is two foundreys at manchester in full work at them and one in Liverpool; there is 6 nearley finishd at Coalbrooke and 7 at a foundry in bridge north begun, and I am makeing Drawings for several more foundreys, any number of them wod sell. There is a vast number of them now errecting and no other engine is errected where these is known. The engine for the West India docks was neglected in my abstance from the Dale but I expect it will be ready to send of[f] in ten days. In abt 3 weeks from this time I shall be in London to set it up; it will please you much for its a verry neat and compleat job and I have no dought but it will answer every purpose exceeding well. At Newcastle I found four engines at Work and 4 more nearly ready, 6 of these was for winding Coals one for lifting water and one grinding Corn; that grinding Corn was a 11 In Cylinder Driveing two pair of 5 feet stones 120 rounds pr Mt ground 150 Winchesters of Wheat in 12 Hours with 12 Hund^d of small Coals. It workd exceeding well and was a very compleat engine only the stroake was much to short onely 2ft 6^{In} Stroake which made it very much against the Duty. The other engine that was lifting Water was a $5\frac{1}{2}$ Inch Cylinder 3 feet stroake Drawing one Hund^d Gall^n barralls 24 every hour 80 yards with 5 Hund^d of Coals in 24 Hours this it did with very great ease. I beleive you will find its exceeding good duty for a $5\frac{1}{2}$ Inch Cylinder.

Below I send you a coppy of Mr Homfrays and Mr Woods letter to mee, Mr Homfrays letter Sep 10th. "Our Engine goes on extreamely well here nothing can go better, the piston gives no trouble goes aboute three Weeks and we work it with black lead & Water and the Cylinder is as bright as a looking glass. It uses aboute two pounds of black lead in a week. Abt once in 12 or 15 Hours we put a small quantity of black lead mixd with a little water through the hole in the Cylinder Cover, and we never use any grease. We rolld last week 140 Tons of Iron with it and it will roll as fast with the both pair of Rolls as the[y] can bring it." Mr Woods letter Sepr 12: "We are going on as it is likley we always shall in the old dog trot way pudling & Rolling from the beginning of the Week till the end of it, Your engine is the favorite engine with every man aboute the place and Mr Homfray says its the best in the kingdom"—I have not the smallest dought but that I can make a piston with out any frecition or any packing whatever, that need not to have the Cylinder Cover taken up on[c]e in 7 years, and its a very simple plan. It will be perfictually tight; its by restoreing an equal liberum on both side the piston. I expect I shall see you in London soon and then I will give you the plan for inspection, before I put it in practice. I am very much obligd to you for recomending these engines in Cornwall but you have not stated in what mannere the[y] are to be applyd, wither to work pumps or Canalls or both. The[y] may be made both winding engines and pumping engines at the same time if required. A rotative engine will cost more than an up and down stroke on account of the expence of the fly wheel and axel. An engine capable of lifting 180 Gall of Water pr Mt 20 fathoms wod cost when compleat and at work on the spot patent right included abt £220—but if its a rotative engine with a winding barrall it will cost £270. I shod expect that a 7 In Cylinder wod be sufficient for winding at Penberthy Crofts which might have a crank on the fly for lifting water in pumps and a winding barrall on its back; this wod cost abt £170. The errection of them when on the spot will cost nothing. You do not say when you intend to be in town; I hope you will be present when the dock engine is sett at Work. The engines that is first sent in to Cornwall must be from Coalbrooke Dale and then the[y] will be well executed but from Wales it wod not be so. You may depend on haveing a real good engine sent down and sufficent openings given to the passages. The engineere from the Dale have been latley at London and just Returned he gives wonderfull account of the engines that is working in London. There is 12 now at Work there the[y] have well establishd there utility in different parts of the kingdom and any number wod sell. The founders intends makeing a great number of different sizes and send them to Different markets for sale complaitley finishd as the[y] stand

you do not say anything abt Wheels to the engine for Pemberthy Crofts. Derect for mee at the Talbot Inn Coalbrooke Dale

I am Sir your very H^b Ser^t

R$_{ICH}$^{D} T$_{REVITHICK}$.

There is several engines here nearley finishd and if the[y] sute in size for penperthy, it may be sent down in 4 or 5 Weeks but otherways it may be 2 Month.

It can be assumed that most of the engines described in the above letter were like the engine of which we give an illustration (Plate XI) which is typical of what Trevithick was building about this time. It was made by Hazledine and Co. of Bridgnorth, but its exact date is not known.

Although the period of Trevithick's early years was one of enormous change in world affairs—we need only instance the American War of Independence and the French Revolution—yet it had little direct influence, as far as one can see, upon his life. The industrialization of this country, which was in essence the task upon which he was engaged, took up all his energies. Indeed the same detachment is apparent in the lives of most of our great inventors, engineers and industrialists at the end of the eighteenth and beginning of the nineteenth century.

However, the danger of invasion of England by the French in 1798 did strike home to every part of the country and the response to the threat was, as everyone knows, a rush to volunteer for home defence. A Cornish detachment of Volunteers was formed by Lord Dedunstanville and in this troop Trevithick enrolled himself. A story is told of him[1] that reminds one of the incident of Sir Francis Drake on Plymouth Hoe when the Spanish Armada hove in sight.

One night a beat of drums in the Camborne streets startled the sleepers; Trevithick awoke his wife, and asked what all the noise could be about. "Oh! I suppose the French must be come; had you not better put on your red coat and go out?" "Well, but, Jane,"

[1] *Life*, I, 326.

PLATE XI. TREVITHICK'S HIGH-PRESSURE ENGINE
AND BOILER BY HAZLEDINE & CO., *c.* 1805

Courtesy of the Science Museum, South Kensington

suggested the volunteer, "you go first and just look out at the window, to see what it is!"

A more serious danger was the menace of invasion of England by Napoleon in 1804. He collected in Boulogne harbour a flotilla of transports and assembled troops there in readiness for a descent on the South Coast. Only the vigilance of the British ships in the Channel frustrated his designs. But besides these watching tactics, active measures were being taken to carry war into the enemy's camp and destroy the flotilla in the harbour.[1]

It was merely by chance that Trevithick was not drawn into the maelstrom of war. The idea was to construct a fire-ship that could be propelled or towed by a steamboat into the midst of the flotilla and then exploded. Had the idea been followed up, the steamboat would have had an earlier application in Great Britain than it actually had.

How Trevithick became concerned in the matter appears from the following letter to Giddy:[2]

Several Gentlemen of late have call'd on mee to know if these engines wod not be good things to go into Bolong to destroy the flotela, &c., in the harbour by fire ships. The[y] told mee that a Gentleman from Bath was then in London trying experiments under Goverment for that purpose, but wither by engines or by what plan I do not know.

There was a gentleman sent to speak to mee yesterday on the buisness, from a marquiss; the name I am not at liberty to give you. I put him off without any encoragement, becase I wod much rather trust to your opponion and bringing this buisness forward than to any other man. I have two 10 In. Cylinders here compleatley ready; they are exceeding well executed, and I will not part with them untill I hear farther from you on the buisness. If you think you could get Goverment to get it putt in to execution, I wod readley go with the engines and risque the enterprize. I shod think that its posiable to make these engines to drive ships in to the middle of the fleet, and then for them to blow up. However, I shall leve all this buisness to your judgement, but if you give mee encoragement respecting the posiability of carrying it in to force, I am ready to send off these two engines on

[1] How real the scare was and what were the schemes proposed or adopted are detailed in Wheeler and Broadley, *Napoleon and the Invasion of England— The story of a Great Terror*, 1908.

[2] *Enys Papers. Life*, I, 321. Letter dated Coalbrookdale, 1804, Oct. 5th.

speculiation. I beleive if you do not bring this buisness forward, some other person will, and it wod not please mee to see another person take this scheme out of our hands. Be silent abt it [at] home, for I shod not like my family to know that I wod engage in such [an] undertakeing. I wod thank you for your sentiments on this buisness soon, as these two engines will be left unsold untill I receive your answer.

The readiness of Trevithick to go "all out" on such a signally dangerous enterprise is quite typical of him, and the request that his family should not be told about it shows his consideration for them.

Giddy replied offering, as he always did, to further Trevithick's projects in every way. The latter had in the meantime been invited by the Marquis of Stafford—the "marquiss" mentioned in the previous letter—to Trentham Hall, near Newcastle-under-Lyme, to discuss possibilities. We infer that his lordship was greatly impressed with Trevithick and his *savoir-faire*; what followed appears from his next letter to Giddy:[1]

I received your last letter, and am very much obliged to you for your goodness in offering to give mee your assistance in promoteing my scheames. I am order'd by the marquis of Stafford to leve this place for London on Monday morning to meet Lord Menville and Mr Pitt. It will be likeley they may wish to be satisfyde respecting the posiability of the plan I shall lay before them, and I shall refer them to you for your oponion and Calculations. I will write You as soon as I have seen these Gents, and will state every particular respecting the buisness. There is a great many things that I wish to comunicate to you, for I cannot satisfy myself from my own figures. I hope that you will be call'd in to the House in the course of next month, and I expect I shall be in town at the same time. If I am order'd to proceed with this plan for government, there will be several things that I shall not be able to get through with out your assistance. I must beg your pardon for so often troubleing you for these things that I am not master of myself, but I hope you will have the pleasure of seeing these plans carry'd in to execution.

What happened further is related in the next letter to Giddy:[2]

---

[1] *Enys Papers. Life*, I, 322. Letter dated Trentham Hall, near Newcastle, 1804, Oct. 20th.

[2] *Enys Papers. Life*, I, 324. Letter dated Soho Foundry, Manchester, 1805, Jan. 10th.

I was sent from the Marquis of Stafford to Lord Menvle [Melville].
I was at the admeraltrie office and was order'd to wait a few days
before the[y] culd say to mee what the[y] wanted. I call[ed] 5 or 6 days
foll'ing and never received a satisfactorey answere, only to still wait
longer. But I left them without knowing what the[y] wanted of mee
for I was tired waiteing, and was wanted much at Coalbrookedale at
the time. When the[y] send for mee again, the[y] shall say what the[y]
want before I will again obey the call.

Typical of his quickness of brain and capacity for improviza-
tion is his account in the same letter of what he did to realize his
ideas as to a steamboat to tow a fire-ship:

There was an engine, a 10 In Cylinder put in to a barge to be
carried to Maccelsfield for a cotton factorey, and I tryd it to work on
board. We had a fly wheel on each side [of the] barge and a crank-
shaft that was thrown across the deck. The wheels had flat boards of
2 ft 2 In long and 14 In Deep, six on each wheel, like an under shut
water-wheel. The extreamity of the wheels went abt 15 Miles pr
hour. The barge was between 60 & 70 Tons burthon. It wod go in
still water abt 7 Miles p. Hour. This was don onely to try what effect
it wod have. As we had all the apperatus of old matls at the Dale, it
cost little or nothing to put it togeather. I think it wod have drove it
much faster with sweeps.

This trial evidently must have been on the Severn or on the
Shropshire Canal and was quite successful enough to give
Trevithick confidence that steam navigation with his engine
was perfectly practicable and that for the moment was all that
he wanted to be quite sure about.

A drawing of a "Design for a steamboat" by "John U.
Rastrick civil engineer" has been preserved.[1] The engine is undoubtedly a Trevithick one and the design tallies very well with
the description except that the barge is larger. J. U. Rastrick
(1780–1856) was a colleague of Trevithick and as we shall see
later made a great number of engines for him, but Rastrick
"never had anything to do with steamboats", so he said in his
evidence before the Liverpool and Manchester Railroad Bill
Committee, 1825. Further the drawing is dated March 27th,
1813, so we must conclude that it is merely a sketch embodying
ideas of Trevithick.

[1] *The Engineer*, II, 1917, p. 81, 1 fig.

Nothing further transpired, but we can conjecture what might have resulted had not Napoleon, faced by the coalition of Russia and Austria, withdrawn his troops in August from the camp at Boulogne to take part in the campaign of Ulm. Nelson's victory at Trafalgar on October 21st, 1805, finally confirmed Britain's naval supremacy.

We now turn to Trevithick's connection with dredging machinery. At the beginning of the nineteenth century much thought was being given to dredgers owing to the vista of possibilities opened out by the steam engine, which was applicable obviously as the motive power. The push-plate conveyer dredger actuated by a current wheel[1] or by horses we owe to Dutch engineers in the seventeenth century. Other types—the bag and spoon, the dipping bucket and the ladder of buckets—were known. Boulton and Watt supplied a bell-crank engine in 1796 to John Grimshaw of Sunderland, who applied it to a bag and spoon dredger. John Rennie in 1802 designed a ladder dredger worked by horses for Perry and Well's Dock at Blackwall and in 1804 applied the steam engine to the bucket dredger.

General Sir Samuel Bentham[2] is to be credited with having in 1800 proposed the application of a steam engine to a bucket dredger; it was built and ready for trial two years later; it was put into service in April, 1802, and during 1803 was almost constantly in use. Owing to its success, Bentham, on June 28th of that year, proposed to build a number of such "digging machines" for H.M. dockyards, but first, as he had "great hopes, that by the adoption of a new kind of steam engine lately introduced by Mr Trevithick, the apparatus may be simplified and rendered less expensive", he merely recommended a larger machine. This and the use of Trevithick's engine were approved; the latter was ordered in 1803 as we know from Andrew Vivian's account book, but the explosion at Greenwich led to the

---

[1] *Trans. Newcomen Soc.* IV, 32.

[2] An account of Bentham's work on dredgers fully documented will be found in *Mech. Mag.* XLIII, 1845, pp. 113–120, by T. G. Chesnel. This is supplemented by the Goodrich papers in the Science Museum, South Kensington. See *Trans. Newcomen Soc.* III, 4.

cancellation of the order. No doubt it was this order from the Navy Board that directed Trevithick's attention to the subject of dredging. The evidence we have as to what he did is none too lucid. What he said himself, in reply to a correspondent who had made an enquiry about dredging in Falmouth Harbour, is as follows:[1]

> I made three engines with machinery for lifting mud at the entrance of the East and West India Docks, and also for deepening the water at the men-of-war's mooring ground at Woolwich. One of these engines was a 20-horse power, erected on an old bomb-ship of about 300 tons burthen, which machine cost (exclusive of the ship) about 1600*l*. This engine would lift and put into barges near 100 tons of mud per hour. Another engine of 10-horse power I erected on board an old gun-brig of about 120 tons burthen, which cost (exclusive of the vessel) about 1000*l*., which lifted about half the quantity of the large one; and another engine of 10-horse power I erected on board a barge of about 80 tons burthen.

We have, too, certain reminiscences and recollections which, as might be expected, exhibit some confusion and require sifting to arrive at the truth.

The first recollection is that of the foreman who was assistant to Mr Hughes, possibly the Mr Deverill mentioned below:[2]

> In the year 1806 I erected a steam engine on board the Ballast lighter Brunswick burthen 60 tons for Messrs Hughes, Bough and Mills, which was employed to deepen the River Thames at the East India Moorings, Blackwell. And in the early part of 1807 my late friend Mr R. Trevithic came on board and looked at the machine, and was afterwards engaged by Messrs Hughes, Bough & Mills to erect a dredging engine on board the Blazer Gun Brig, a vessel about 5 times the burthen of the Brunswick. When it was finished, or as it was then thought to be, Mr Hughes, the late Mr De Vaux, Civil Engineer, of the City Canal, removing the rock at Blackwall, &c., &c., and Mr Mills came to me at the East India Moorings, and informed me that the Blazer dredging engine was deficient in some parts of its machinery as it could not perform what it was intended to do. They

---

[1] *Life*, I 248. Letter dated Camborne, 1813, Feb. 4th.

[2] *Enys Papers*. Autograph Letter, Thomas Hughes to J. S. Enys, 1839, Aug. 10th.

then wished me to go on board the Blazer and examine it, but I strongly objected, saying I did not like to interfere with any other person's business. Mr Hughes then told me that if I would not go on board and see what was wrong, the vessel must be put into Perry & Wells' Dock, for it was a very great expense to keep her on the river with two sets of men, without doing scarcely any work at all. Mr Hughes then showed me a calculation of the expense of the Blazer dredging engine and pressed me very much to go on board and make the necessary alteration, and I then complied with his request. The Blazer was after that employed to assist in finishing the contract at the East India Moorings and afterwards to raise gravel near Westminster Bridge, and various other places in the river & docks.

Mr Trevithic was also engaged by Messrs Hughes, Bough and Mills at the same time to erect a dredging engine on board the Plymouth Bomb Brig, lying at Limehouse hole: but when Hughes and Co. found the Blazer dredging engine defective, they requested me to go on board the Plymouth & make a report to them on the subject, which I did, and then the Boiler, Engine, Framing etc., were taken out of the vessel immediately. Messrs Hunter and English of Bow, Middlesex, wheelwrights, were then employed to erect a dredging machine on board the Plymouth, which was first set to work at Woolwich Dockyard to good effect.

The late Mr John Rennie and the late Mr Watt came on board the Brunswick in 1807, and afterwards built a dredging engine to drain the Fens in Lincolnshire.

The second recollection on the subject of dredgers is that of Thomas Bendy, a millwright who had actually worked for Trevithick on one of the dredgers. In 1828 he gave Francis Trevithick some information and supplemented it in 1840 by writing as follows:[1]

about the dredger engine and machinery made by your father for Mr Bough. It was fixed in the year 1803 and was altered by Mr Deverill...in 1805. The cylinder was 14½ inches in diameter, the stroke 4 ft, the chain ladder 28 or 30 feet. The largest quantity of stone and gravel lifted in one tide was 180 tons. The reason for using the word stone is from its being part of Blackwall rock. The engine was cast at Hazledine's at Bridgenorth but finished at Mr Rowley's factory in London[2] by some men from Cornwall, and a part of the

---

[1] *Life*, I, 240.
[2] This was at 7, Cleveland Street, Fitzroy Square.

machinery by Jackson, a Scotch millwright. The working time between tides was from six to eight hours in from 14 to 18 ft. of water at the entrance to the East India Dock. The expense of the engine and machinery [was] a little more than £2000. I should think the time she worked was about ten or eleven years.

The other engine you mentioned was the property of the Trinity Company and Government, by whom made I cannot ascertain. It was worked in the river near Woolwich entrance dock gates but not to much account only lifting mud to clear the entrance. No new materials were taken there by Mr Deverill; all that was done was to refix the old and repair the engine. This engine was the same size as that fixed in London to run round the circle at the speed of fourteen or fifteen miles per hour....

We have a third reminiscence, in 1869—that of Captain John Vivian[1] whom we have already mentioned as having driven the London road carriage of 1803. He said that in that year he "saw Trevithick breaking the rock at the East India Dock entrance to the Thames at Blackwall using a water-wheel worked by the tide and also a small high-pressure engine for driving or turning large chisels and borers and other contrivances for breaking and clearing away the rock".

We have here two distinct episodes, viz. excavating the tough boulder clay at the new Blackwall entrance to the East India Docks, and dredging mud, etc., in the Thames estuary.

As regards the apparatus used at Blackwall Mr Joseph Glynn stated[2] that it consisted of a weight working in guides armed with a steel chisel to break up the boulders; it was used in conjunction with a cylindrical iron coffer-dam out of which the water was pumped. This apparatus, he states, was made at Butterley Iron Works by William Jessop (1745–1814). We can only surmise, therefore, what, if anything, Trevithick had to do with the excavator for the entrance to the Dock.

About the dredgers for raising mud or ballast, however, we have quite a considerable amount of information. There cannot be much doubt that one, at any rate, of the dredgers mentioned was that constructed for the service of the Trinity House for

[1] *Life*, I, 240.
[2] *Mech. Mag.* XXXIX, 1843, p. 309.

raising ballast to supply the shipping resorting to the Thames. Fortunately we have in the minute books[1] a full account of what took place.

It is necessary to turn back a little to observe that the Corporation of the Trinity House had promoted a private Act of Parliament, 45 George III, cap. xcviii, 1805 "...for the Regulation of Lastage and Ballastage in the River *Thames*; and to make more effectual Regulations thereto". The Act recites the privileges that the Trinity House enjoyed, 36 Elizabeth and 17 Charles II, to take ballast from the river and supply it to shippers, repeals subsequent Acts 6 and 32 George II and consolidates the whole. One of the provisions was in regard to the cost of heaving the ballast. The ballast heavers were claiming an increased rate on account of the increased cost of living. Consequently "not less than the Sum of Sixpence nor more than the Sum of Eight Pence, at the discretion of the said Corporation" was allowed. Apparently in Committee of the House on the Bill, the shipping interest had suggested that the heaving of ballast might be done by steam and so avoid increased burden on trade. Hence the Corporation advertised publicly for tenders for heaving ballast. The archives of the Trinity House show that Sir William Curtis, Bart., M.P., Lord Mayor of London, whom Homfray apparently knew, had mentioned at the time of passing of the Ballast Act that Homfray and Trevithick were anxious to make an offer to use the latter's engine for raising ballast. Accordingly they were invited to do so, and the course of the negotiations is very clear from the minute book. On January 2nd, 1806, "a Proposal of Mr Richard Trevethick for raising Ballast by a new Method at 6d pr Ton was read". Trevithick communicated the substance of this to Giddy:[2]

I am aboute to enter in to a contract with the Trinity Board for lifting up the ballast oute of the bottom of the Thames for all the shipping. The first quantity stated was 300,000 Tons pr Year but now the[y] state 500,000 Tons pr Year. I am to do nothing but wind

---

[1] Our abstract has been made from "By-minutes 1804–1808", access to which has been obligingly given by the Secretary to the Trinity House.

[2] *Enys Papers. Life*, ii, 143. Letter dated Camborne, 1806, Feb. 18th.

up the chain for 6d pr Ton which is now don by men. The[y] never lift it above 25 feet high. A man will now get up 10 Ton for 7s. My engine at Dolcoath have lifted above one hund^d Tons that height with one bushell of Coals. I have two engines all ready finishd for the purpose, and shall be in town in abt 15 Days for to sett them at Work.

Either Trevithick had not realized what the conditions really were or he was as usual over-sanguine. On February 20th "It was RESOLVED That all such plans be referred to John Rennie, Esqr., Engineer, to examine and consider and report his opinion thereon". He was also to be asked "whether he can himself devise any Plan or Method by which Ballast may be raised and shipped to the Shipping from the Bed of the River Thames, at a cheaper Rate than by the Method now in use": i.e. manual labour with the spoon dredger.

Trevithick attended on the Supervisors on May 22nd, 1806, "and explained the Outlines of his Plan for raising Ballast by a Steam Engine of his Own Invention". This was found "in many Respects objectionable" and he promptly submitted "a Second Proposal". Thereupon an agreement for twenty-one years, modified to fourteen years, was drawn up, but the Board decided, wisely: "it will be most expedient not to enter into any Agreement until the proposed Experiment has been found successful".

Trevithick got to work as he had promised, and mentions progress in a letter to Mr Giddy,[1] according to whom Trevithick had much trouble with breaking of the chain and buckets:

This Day I set the engine at work on board the Blazer gun brigg. It does its work exceeding well. We are yet in dock and have lifted up mud onely. I hope to be down at Barking Shelf in a few Days at our proposed station....I think theres no doubt of sucksess.

There is an illustration of a dredger in Rees's *Cyclopedia*[2] which obviously answers to this description: the engine is undoubtedly a Trevithick one and the bucket ladder is situated at the side, just as it would be in an improvised job (see Fig. 10).

[1] *Enys Papers. Life*, I, 327. Letter dated Black Wall, 1806, July 23rd.
[2] Plate III, "Hydraulics", no scale given: dated 1812.

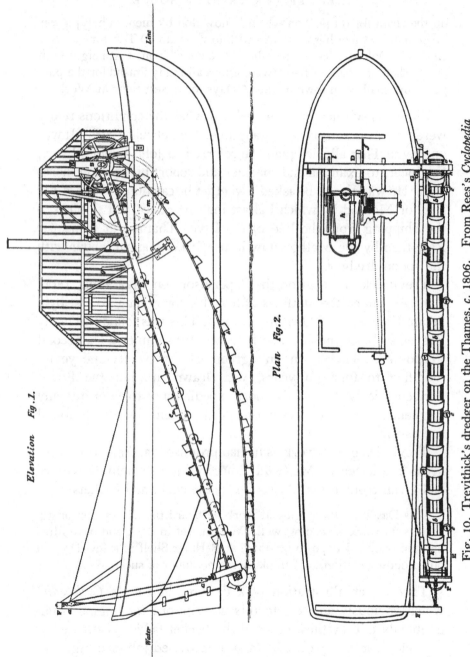

Elevation Fig. 1.

Plan Fig. 2.

Fig. 10. Trevithick's dredger on the Thames, *c.* 1806. From Rees's *Cyclopedia*

In the article "Dredger" a description is given but Trevithick's name is not mentioned. The engine is stated to be of 6 horse-power and to be capable of loading a small barge with ballast in $1\frac{1}{2}$ hours. The authors are satisfied that this illustrates Trevithick's dredger because in the article "Steam Engine" the "high pressure Steam Engine used with the Dredging Machine on the River Thames" (see Fig. 11) is credited to Trevithick and is so obviously his.

On November 27th, 1806,

The Supervisors of the Ballast Office acquainted the Board that Mr Trevithick has removed the Vessel with his new invented Steam Engine and machinery for raising Ballast, from Barking Shelf (where she has been frequently run on board) to Limehouse Hole; and has made application for Lighters to be sent him at the latter Station, to receive the Ballast he proposes to raise there.

This did not quite suit the Board; they were beginning to be restive and sent Trevithick a letter "desiring to know what farther Time will in his own Opinion, be necessary for the Purpose of ascertaining the Utility of his Machinery".

On December 11th, 1806,

a letter from Mr Richard Trevithick dated this Day was read, detailing the Result of his Experiments with his new invented Engine for raising Ballast and signifying that finding the present Machine not sufficiently powerful, he proposes to erect One of Six Times the Power thereof which will cost him Five Thousand Pounds in addition to Two Thousand already expended; but still he is ready to go forward provided the Corporation will grant him an Increase of Price of Three Pence pr Ton on all Ballast, deliver'd by the machine on board Ships; he therefore requests to know if he should take out the small Engine, and put his large one into the Vessel? or whether the Corporation are inclined to sell the Vessel, or would purchase his small Engine now on board her?

Obviously this proposal would have defeated the very purpose the Corporation had in view, i.e. to keep down the price, and in fact the Ballast Act did not permit them to go beyond 8*d*. per ton.

The Corporation, having got a proposal from Messrs Hughes,

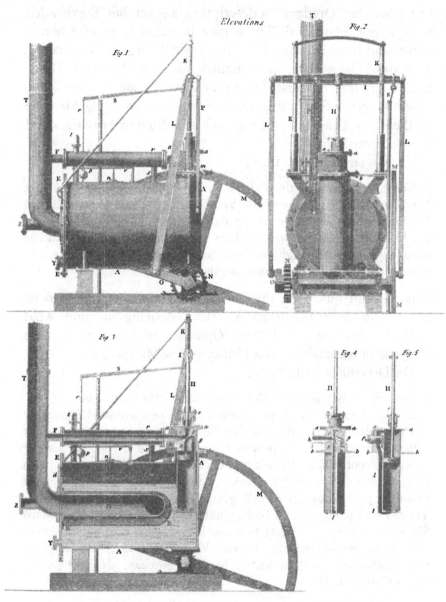

Fig. 11. Trevithick's high-pressure engine used with the dredger on the Thames, *c.* 1806. From Rees's *Cyclopedia*

Bough and Mills for raising the ballast and putting it into lighters by the old method for 7½*d*. per ton, told Trevithick they could not give him more than that sum, especially "as Experience hitherto has not proved it [i.e. Trevithick's machinery] capable of delivering any Part of the daily Supply with the Regularity and Dispatch required"; at the same time they agreed "to sell him the Ballast Brig for £300, the Price he has offered for the same".

After being dismissed by the Trinity House, Trevithick seems to have induced someone to join him in purchasing the dredger, for in a letter to Giddy on January 5th, 1808, he says: "I have sold the ballast business to a company which is carrying it on". We may hazard the guess that Messrs Hughes, Bough and Mills were the purchasers for contracting work, and on the testimony of Bendy the dredger continued to be used in the Thames for ten or eleven years. We conclude therefore that the "three engines with machinery" that Trevithick claims to have made were for the "Brunswick" ballast lighter, 60 tons, the credit for which is disputed; the "Blazer" gun-brig, 120 tons; and the "Plymouth" bomb-vessel, 300 tons, which was apparently a failure. In every case we take it that Trevithick's principal share was the adaptation of his high-pressure engine to the dredging machinery.

The method of raising ballast by manual labour continued in use for many a long day. No move was made when the matter was reopened in 1824 by the Shipowners' Society, who suggested the desirability of "procuring Ballast in the River by the employment of Steam Engines": the supervisors to the Corporation in a résumé of the matter stated:

they gave Mr Trevithick a fair Trial but finding that he could not dispatch more than four Lighters in one Tide, and that the project was totally inadequate to the object in question, this scheme after many communications naturally fell to the ground as well as that of the other projector [i.e. Edward Shorter].[1]

However, circumstances had now changed and the result of

----

[1] Letter, 1824, May 14th, preserved at the Trinity House.

this agitation was that the Trinity House in 1826 found that
it would be an economy to employ a dredger of their own.

As if he had not enough irons in the fire already Trevithick
must now interest himself, owing to the recommendation of
Giddy, in a project for tunnelling under the river Thames.
Mining enterprises necessitate much driving of levels or adits
and the methods were well understood, but subaqueous tunnelling with the ever-imminent danger of the inrush of water was
then quite a novel undertaking. The idea of a tunnel had
arisen owing to the need for greater facilities, than were afforded
by the existing ferries, for crossing the Thames below London
Bridge. This was not the first proposal of its kind but it was
supported by prominent men in the City. Robert Vazie, an experienced Cornish engineer, was selected to carry out the work,
indeed he seems to have been the original promoter. He made
borings on both sides of the river and reported "that the work
would not be so expensive as had been expected". What had
been the anticipated expense we do not know. Subscriptions
were raised from a number of gentlemen who obtained a private
Act of Parliament, 45 George III, cap. 117, "for making and
maintaining an archway or archways under the River *Thames*
from the Parish of *Saint Mary Rotherhithe*, in the County of
Surrey, to the opposite Side of the said River in the County of
Middlesex" (July 12th, 1805). These gentlemen were constituted
"One Body Politick and Corporate, by the Name and Style of
the Thames Archway Company". The tunnel was to be "pass-
able for Horses, and Cattle, with or without Carriages, and for
Foot Passengers". The capital was £140,000 in shares of
£100 each, with powers to raise £60,000 on mortgage. The
names of Vazie and several members of his family appear in the
Act as shareholders. The first meeting of the proprietors was
held on July 25th, 1805, at the Globe Tavern, Fleet Street.

Vazie's plan was to sink close to the bank of the river and,
when he had got sufficiently deep, to drive a trial level or drift-
way on the proposed line of the tunnel with the intention eventually of using this as a drain whereby to construct the actual
tunnel or tunnels. Vazie commenced work by sinking a shaft

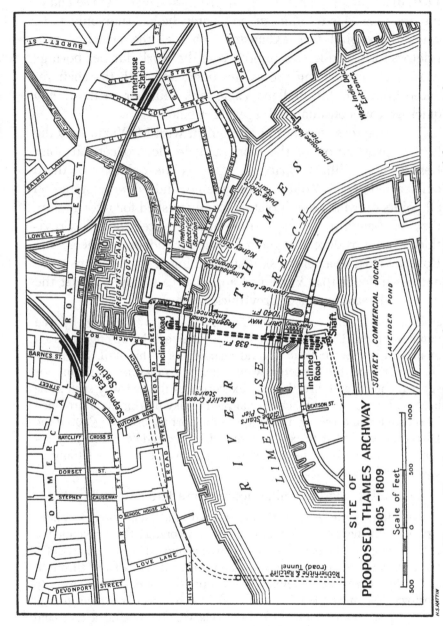

SITE OF
PROPOSED THAMES ARCHWAY
1805 – 1809

Scale of Feet

500   0   500   1000

Fig. 12. Site of proposed Thames Archway, 1805–1809, on a map of the present day

11 ft. diam. at a distance of 315 ft. from the river. At the end of the first year, due to influx of land water, he had sunk only 42 ft. when the capital was exhausted. One of the principal proprietors furnished means to prosecute the work, further borings were made confirming the earlier ones, and with the shaft reduced in size to 8 ft. in diam. Vazie sank a further 34 ft. when a quicksand was encountered a few feet below.

The Directors were not in agreement with Mr Vazie as to the further prosecution of the work and decided to consult John Rennie and William Chapman. But as the opinions of these eminent engineers "did not coincide nor indeed were stated on all points on which the Directors chiefly wished for information, they felt bound to resort to some other source; and Mr Trevithick was introduced to them by their resident engineer, Mr Robert Vazie, as a person skilled in mining". There seems to be not much doubt that Vazie had appealed to Giddy and that the latter had recommended Trevithick strongly:

after a due examination into his character as appearing by the minutes of the Directors, and having received the strongest testimonies in his favour from several quarters as to his skill, ingenuity and experience, the Directors were induced to contract with him for superintending and directing the execution of the driftway such as he proposed it to be; for which they agreed to pay him £1,000 provided he succeeded in carrying it through to the north shore; or £500 if the Directors ordered it to be discontinued in the middle which they reserved to themselves the power of doing; but to receive nothing in case he did not succeed.[1]

Trevithick gives a similar account in a letter to Giddy:[2]

Last monday [i.e. Aug. 10th] I closed with the Tunnel gents. I have agreed with them to give them advice and conduct the Driveing the level through to the opposite side, as was proposed when you attended the commette; to receive £500 when the Drift is halfway through, and £500 more when its holed on the opposite side. I have wrote to Cornwall for more men for them. Its intended to put 3 men in each core [of] six hours' course. I think this will be making a thousand pounds very easey, and withoute any risque of a loss on my side, and

---

[1] Quoted in the *Life*, I, 249, from an official source.
[2] *Enys Papers*. *Life*, I, 252. Letter dated Limehouse, 1807, Aug. 11th.

as I must be allways near the spot to attend to the engines on the river, an hour's attendance every day on the tunnel will be of little or no inconvenience to me. [I] hope 9 months will compleat it. From the recommendation you gave me, they are in great hopes that the job will now be accomplishd; and as far as Capn Hodge and my self culd judg from the ground in the bottom of the pit, theres no dought of compleating it speedily. I am very much obliged to you for throwing this job in my way, and shall strickley attend to it, both for our credit as well as my own profit.

Trevithick, as usual, did not let the grass grow under his feet. In a week's time he had started work, probably with the help of Vazie's men pending the arrival of other skilled men from Cornwall. Trevithick reports[1] progress to Giddy as follows:

Tuesday last was a week [i.e. Tuesday, August 18th] since we began to Drive one level at the bottom of the engine shaft at the archway. The level is 5 ft high, 3 ft wide at the bottom, and 2 ft 6 In Wide at the top, within the timber. The first week we drove 22 feet. This week I hope we shall drive and timber 10 fathoms. As soon as the railway is laid I hope to make good 12 fathoms a week. The disstance we have to drive is abt 188 fathoms. The ground is sand and gravel and stands exceeding well, except its when we hole into leareys, and holeing to such houses of water makes the sand very quick. We have discovered three of these holes which contained about 20 square yards. Its very strange that such spaces shod be in sand at this Deypth. When we cutt in to the places we are obliged to timber it up close untill the sand is draynded of the water, otherways it wod run back and fill the drift and shaft. I cannot see any obstickle likley to prevent us from carrying this level across the river in six month as the Engine throws down a sufficient quantity of air, and the railway underground will inable us to bring back the stuff, so as to keep the level quite clear, and the last fathom will be as speedily drove as the first. Theres scarsley any water in the level—not above 20 Gallns pr. Minute—and not a drop falling from the back of the level. Therefore, I think we may expect that the land springs will not trouble us. The spring thats comming down round the outeside of the walling of the shaft is rather encreased. The Directors is in wonderful spirits and everything goes on very easy and pleasant. The engines on the river goes on as usual.

[1] *Enys Papers. Life*, i, 253. Letter dated Plough Inn, Kidney Stairs, Limehouse, 1807, Aug. 28th.

Trevithick adds the following postscript:

The two Inch Iron air pipes that was provided before I tuck it in hand is too small. The smith's bellows have near two hund^d weigh on the top plank, and at the bottom of the shaft it will scarsley blow oute a candle. [I] shall put down larger ones next week.

The tunnelling work was one to which all the miners were accustomed; by poling boards outside the timbers they kept out the sand and gravel, and by a tramroad they were enabled to get the spoil away quickly.

We have another report in the following month:[1]

Last week we drove and secured in the Tunnel 25 Yards, and I cannot see any dought of getting on in future with the same speed. We are now abt 180 feet from the shaft, and as we approch the river the ground is better, and the water do not increase; but to be proofe against the worst, its agreed to have a 30 Horse power [engine] in readyness to assist in case of cutting more water than the present engine can cope with. The Distance from the shaft to the spot on the opposite shore, where we intend to come up to the Day, is about 1220 feet. This disstance I hope to accomplish in a short time, unless an unfavourable circumstance turn up, which at this time do not show itself. If the ground shod prove softer, we can Drive horizontal piles, and shod the water increase, we shall have three times the power in reserve that we now occuipy in a very short time. The Directors is highley pleased with our present proceedings, and I have no dought but what we shall continue to give them entire satisfaction. In conciquence of this job I have been call'd on to take the Deriection of a very extencive work, the nature of which I am not yet fully inform'd; but am to meet the party on Thursday next for farther informeation. As soon as I am fully in pocession of this plan I shall comunicate its contents to you for your investigation; at the same time [I] must beg your pardon for so often troubleing you on matters that cannot advantage you, but at the same time hope you will excuse my freedom, being driven to you as a sourse for informeation that I cannot be furnish'd with from any other quarter. The engines continue to get on as usual on the river. The great engines is not yet at work, but hope it will soon be compleated.

[1] *Enys Papers. Life*, 1, 254. Letter dated Plough Inn, Limehouse, 1807, Sept. 12th.

Continuing our account based on the official report:

On the 5th of September following, Messrs. Vazie and Trevithick, in a joint report to the directors, strongly recommended the immediate purchase of a 30-horse-power steam-engine. The directors did accordingly purchase the same, and it is now ready to work. The driftway proceeded till about the beginning of October, when it appeared that the works had been very considerably interrupted and delayed in their progress. The directors therefore, on the 8th October, resolved to institute an inquiry into the cause; and the consequence of this investigation and disclosure of facts was the removal of Mr Robert Vazie from his office as resident engineer, on the 19th October, by which time the drift had been extended 394 feet, that is, at the rate of 6 feet 2 inches per day, through a dry sand.

We can only conclude from this report that Vazie and Trevithick had not got on well together—which is highly probable; indeed it was an anomalous situation to have two executive men on the job. Vazie had to go, but it was hard on him for he had spent $4\frac{1}{2}$ years on the work, during which he stated he had not slept one night away from the work.

The works now proceeded without embarrassment, and with considerable less cost; as from this time (the 19th October) to the 29th November, the ground continuing as nearly as possible of the same quality, it was extended 421 feet, or 11 feet 2 inches per day, which is nearly a double rate (deducting three days and a quarter that the works were suspended while the directors determined on the turn the drift should take).

From the 29th November to the 19th December the drift was extended only 138 feet, or 6 feet 10 inches per day, in consequence of the drift now running in a stratum of rock, great part of which was so hard that it could not be broken up without the use of chisels and wedges.

By the 21st December the drift had proceeded 947 feet from the shaft; and it was observed that the strata through which it passed dipped to the northward about 1 foot in 50, in consequence of which the rock that at one time formed the whole face of the breast, now only reached within 2 feet of the top, and which was occupied by a sandy clay, mixed with oyster and other shells, and containing some water.

On the 23rd, notwithstanding the workmen were proceeding with the utmost precaution, the roof broke down and discharged a great quantity of water from a quicksand, which was afterwards ascertained to be about 5 feet 6 inches above the roof.

By the 26th January the drift was extended 1028 feet, having been worked through a considerable part of the quicksand; and at this period the river made its way into the drift by a fall of earth, which made a considerable orifice in the bed of the river, which has been filled up at several times with earth, carried there for that purpose, and the drift has since then been extended to 1040 feet, which is the present length of it.

The discharge of Vazie had given umbrage it seems to several of the proprietors, one of whom especially left no stone un-turned to decry Trevithick's moral and professional character, undermine the confidence of the other proprietors in him, and interfere with him as much as he could. The progress of the work in the face of these hindrances is alluded to in a letter to Giddy:[1]

I shod have wrote you on the receipt of your letter, but at the time it arrivd we were in a quick sand, and I wishd to have goat through it before I wrote you. The Drift is drove 952 feet from the shaft and is now aboute 140 feet from the high water mark on the North side [of] the river which I hope to have through in abt a fortnight. Some weeks we have drove 20 fathoms, but for 15 or 20 days before we meet the quick sand we had a very hard lime rock which very much impeded us, and the last fortnight scarsley anything have been don in the end on account of the quick sand; but now we are again in a strong clay ground, and getting on fast. The Drift is 72 feet below high water mark, and when we began to drive from the shaft it was a firm green sand; soon a bead of gravel of abt 6 In thick came down from the back, and as we drove north we find all the stratums dip aboute one foot in fifty. On the top of this gravel there was abt 3 feet of clay, and then a limestone rock made from water of about 5 ft thick. Its evident that this is a made rock, becase the green sand that we began to drive in is several feet under the rock, and there is a great Quantity of bows of trees in it petered in to a stone. On the top of the rock their is a proper bead of oysters, mud and oyosters abt 2½ ft thick. Above this is abt 5 ft of clay mixd with sand; above that is the quick sand of 2½ ft thick; and above that is a strong dry clay which we suppose holds up to the bottom of the river which is abt 20 feet. We bored on the north side and found this quick sand was above the back of the drift, and a great Quantity of water in it. I proposd boreing up in the back of the

[1] *Enys Papers. Life*, i, p. 259. Letter dated Thames Archway, 1808, Jan. 5th.

drift to tap it and let down the water gently through an iron pipe, but I culd not be premitted. In course of working in the drift the water and quick sand broake down the back of the Drift, and drowned the engine, and threw a great Quantity of sand down in the Drift. This sand is nearly as fine as flour, and when in water [is] exceeding Quick, but we had the good fortune to stop up the drift tight with timber before the water goat up to prevent us, and we have drained the sand, so that the engine is again compleat master of it. I had a machine made to drive very long flat Iron bars close up to the back of the Drift, through the stopping boards in the end; and then clear oute the Drift under them, and timber verry close under the bars.... By driving a great number of bars of Iron I again goat to the end of the drift, and its now in good course of driveing again. After we had stoppd the end of the drift to prevent the quick sand from comming back, I bored 14 feet up in the back of the level, and put up a two-Inch iron pipe above the Quick sand, which prevented the quick sand from comming away with the water, and also tuck off the weight of the pressure of water from the broaken ground in the back of the Drift. It was strongley proposd by one of the properitors when the drift was half-way in to open from that place the tunnel to 16 ft high & 16 ft wide. I refusd to do it, knowing [that] this water and quick sand was over our head, and that as soon as we began to incline the bottom of the Drift to the surface on the north side we shod be into it. It was with the greatest difficulty that we culd stop it in the drift, onely 2½ ft wide and 5 ft high, and if we had opened the Tunnel to the full size, every man that might have been underground at the time must have been lost, and the river through to the Tunnel in 10 Minutes, for the water wod have brought the 2½ ft stratum of quick sand into the Tunnel, and then the clay roofe wod have sunk under the weight of the river; for it wod have been impossible to have stopd it in to 16 ft high & 16 ft wide; and the engine would have been dround in one minute, and the sand wod have con- staintly com away under water untill the roofe fell through to the river. This properitor have been very much exasperated against me ever since, because I wod not open the tunnel from the middle of the drift up to the full size. This gent was never in a mine in his life, neither do he know any thing about it. He calld a general meeting to disscharge me, but he was taken no notice of, and the thanks of the meeting given to me for my good conduct, and his friend Mr Vaize disscharged. They have offered me £1250 more for my attendance to open the Drift up to the full size of the Tunnel, and wish me to engage with them immidtly, before the first contract expires. The quick sand when draind is very hard, and after the drift is through the ground will

be so compleatly draind that I cannot see any risk in opening to the
full size. When you come to town I shall have several things to lay
before you, and be very much obliged to you to say when you expect
[to] be hear.

It now seemed as if Trevithick was going to be permanently
stationed in London and his visits to his wife and family in
Camborne became so few and far between that early in 1808 he
persuaded her to join him in London. Their son says:[1]

There had been much correspondence about the wisdom of this
move. Mrs Trevithick's brother, Mr Henry Harvey, advised her not
to leave her home and friends, until things were more settled and
more certain in London. Trevithick's notes to his wife, however,
made everything easy and agreeable. More than 300 miles had to be
travelled in a post-chaise occupied by herself and her four little ones;
the youngest of them a baby. The contrast between her clean and
fresh Cornish home and the habitation at Rotherhithe did not help to
remove the fatigue of the journey, and a further disappointment
awaited her. In her husband's pockets were two of her last letters
unopened. What reasons could possibly be offered for such hard-
hearted ingratitude? Trevithick's answer to the charge was simply,
"You know, Jane, that your notes were full of reasons for not coming
to London, and I had not the heart to read any more of them".

An ingenious and ready answer but not one calculated to
"turn away wrath". Francis Trevithick says the house was
"in a dingy situation near the mouth of the driftway" and that
on the latter closing down, the family removed to 72, Fore Street,
Limehouse, but as Trevithick began making riveted tanks there
—of which more anon—"this noisy residence could not have
been much more agreable than gloomy Rotherhithe".

During these trials Mrs Trevithick had the consolation of making
the acquaintance of a friend in adversity. Mr Vigurs, a Cornish
acquaintance, had married a lady, driven by the French revolution from
luxury in her native land to comparative poverty in London. The two
ladies consoled one another over a cup of tea and a Cornish pasty at
Limehouse; and on the return visit to Bond Street by a sample of French
cookery.[2]

To come back to the drift-way. The last break-in of water,
January 26th, 1808, was considerably helped by a high tide on

[1] *Life*, i, 292.   [2] *Life*, i, 293.

that day. The water rushed in in such volume that the men managed to get away with only breathing space between the surface of the water and the top of the drift. Trevithick was in the drift at the time, and being the last to leave, was nearly drowned. His wife mentions him coming back home to her through the streets, covered with clay, without hat or shoes, but nevertheless undaunted. Under the best circumstances, in the confined space of the drift-way where a man could not stand erect, pass another without squeezing, or work more than one at a time, with water constantly dripping over him, great perseverance was called for. Yet this huge break-in was mastered by the device of throwing in clay over the hole, draining out the water from the drift-way, driving flat iron bars between the stoping or poling boards, and tapping the water in the quicksand met with by 2-inch iron pipes. Trevithick invented a machine to do this—very typical of the resource of the man—and in this way 12 ft. more was got out. In a letter to Giddy[1] Trevithick refers to this extraordinary mental and bodily toil as if it was quite an everyday occurrence:

Last week the water broake down on us from the river, through a quick sand, and fill'd the whole of the level and shaft in 10 minutes, but I have stop'd it compleatley tight, and the miners [are] at work again. We are beyond low water Mark on the north side with the drift; and if we have no farther delays we shall hole up to the surface in 10 or 12 days. I cannot give you as full a description, as I culd wish by the pen; but I will see you on Friday morning. On Thursday, [i.e. Feb 4th] 12 o'clock, there will be a meeting of the Directors, on the spot. and If they new that you was in town, you would be press'd very hard to attend. If it suted you to attend I sho'd be verry happy to see you on the spot as I have a great deal to communicate to you. I have no dought of accomplishing my job.

On February 4th, 1808, the meeting of the Directors took place on the site. Trevithick put forward a plan for dealing with the situation;[2] this consisted, briefly, in making a coffer dam of

[1] *Enys Papers. Life*, i, 264. Letter dated Thames Archway, 1808, Feb. 2nd.

[2] See *Life*, i, facing p. 266, plate copied from his original drawing.

sheet piling on the site of the break in the bed of the river with a shaft in the centre carried down to the roof of the drift-way. The annular space was to be filled with clay. Another shaft near the shore was to be sunk through the bed of the river to facilitate removal of spoil. The whole matter was considered at a meeting of the proprietors on February 29th and it was decided to obtain the advice of two more engineers. Meanwhile Trevithick, ever the optimist, struggled on, as is shown by his letter to Giddy:[1]

We are still following the same plan at this place as when I saw you at the Brewery [i.e. Meux's] but from the present appearances, I expect we shall be through the quick sands in a few Days and hope that 8 or 10 days from that time will accomplish and bring the Drift to Daylight on the opposite shoore.

What the Directors did is best told in the words of the circular[2] issued in April, 1808, to the proprietors:

Since the removal of Robert Vazie from the office of resident engineer to the Thames Archway Company, several proprietors in that concern have expressed their dissatisfaction at the proceedings of the works under Mr Trevithick, and have taken many steps to impede their progress and darken Mr Trevithick's reputation; yet, as no act of Mr Trevithick's incompetency was ever shown, though many were falsely alleged, and as the directors never observed any instance either of neglect or want of skill in him, but that on every occasion where his knowledge, his intelligence, and experience in his profession were questioned and examined by competent persons, his talents appeared very superior to the common level, the confidence which the directors reposed in him was not shaken. But the directors, upon the suggestion of the dissatisfied proprietors, and with a view to gratify their wishes, consented on the 29th February, 1808, to apply to two professional miners of high reputation in the North of England, approved by themselves, to come and examine the state of the works. These gentlemen, namely, Mr William Stobart, of Lumley Park, Durham, and Mr John Buddle, of Walls' End, Newcastle-upon-Tyne, arrived in London on the 5th instant, [i.e. of April] and inspected the works on the 7th; and on that day attended a meeting of the Directors, when the written questions hereafter stated were delivered to them for their consideration and answers; and the meeting of the directors being adjourned to

---

[1] *Enys Papers*. Letter dated Thames Archway, 1808, March 13th.
[2] *Life*, I, 267.

the 9th instant to receive the answers, those gentlemen attended on that day, and delivered to the Board their answers in writing.

It would be tedious to give the questions and answers at length; suffice it to say that these two colliery engineers, men of great experience, approved of every expedient that Trevithick had adopted. We quote one answer:

He has shown most extraordinary skill and ingenuity in passing the quicksand; and we do not know any practical miner that we think more competent to the task than he is. We judge from the work itself, and until this occasion of viewing the work, we did not know Mr Trevithick.

Not being able to throw blame on Trevithick, the Directors, on April 19th, sought advice from a Mr Charlton with no better result; indeed part of what he said was nonsense. As a matter of fact there was no experience to go upon and it was useless calling in one engineer after another, merely to repeat what previously had been said but without suggesting anything fresh.

We have been unable to discover whether Trevithick was allowed to proceed with the caisson as approved by the engineers Stobart and Buddle; possibly there may have arisen difficulties with the Corporation of London from the obstruction to navigation that such a caisson would occasion. During May and June a small amount of boring and piling went on under John Rastrick as resident engineer.[1] The situation was rapidly approaching an impasse. Trevithick, who seems to have been the only one with any constructive ideas, had by this time evidently come to the conclusion that the best way would be to start afresh altogether, make the tunnel in sections and deposit them in a trench excavated in the bed of the river. This was a bold expedient and one, it will be recalled, which has been used in recent years with success.

In order to get the opinion, which he valued so highly, of Giddy, Trevithick wrote him as follows:[2]

The Tunnel is at a stand and a speshall meeting is calld for Saturday

---

[1] *Life*, i, 270.

[2] *Enys Papers. Life*, i, 273. Letter dated Thames Archway, Rotherhithe, 1808, July 28th.

[i.e. July 30th]. Its in conciquence of my proposeing a new method of carrying it on; and if carryd in to effect, which I have no dought will be don, the drift is far enough, and culd not be of any more service if through. The plan I have laid before them is to make a casseoon, 50 ft long and 30 ft wide, and as high as from the bottom of the river to high water, made of whole Balk and the joints calkd tight as a ship. On this casseoon place an engine for Driveing piles and draw them again, and also for lifting the brick oute of the barges, and lifting the stuff oute of the casseoon in to the barges, by a crane worked by the engine. The piles must be drove within on every square of this cassoone and as deep as the bottom of the tunnel. Then take off the earth and make 50 feet long of Brick Tunnel; and then remove the casseoon and draw the piles and fix them 50 ft farther on in the river, and add another pice to the tunnel, untill the whole is finishd. By which plan onely 50 ft of the river would be occipd at one time, which is not as large as a 400 ton ship lying at anchor in the river. We must bore up from the roofe of the Drift and put up a pipe to tap down the water from the Casseon every time we move it. This casseon and engine, with all its matls, will flote all togeather, and allmost all the work will be don by the steam engine. The first plan was a Tunnel of 11 ft Diamr, for foot passengers onely, and 14 feet lower than this plan. This plan is two Tunnels side by side, of 12 ft Diar, each to have a waggon road of 8 ft wide and a foot path of 4 ft wide; and in one Tunnel to admitt [persons going] forward and the other backward, so as to prevent mistook in passing.

This plan can be compleated for £10,000—less than the foot path of 11 feet wide and I think a safe and shure plan, where the other would be [a] very expencive, dangerious, and uncertain plan; and if ever executed would then onely be a foot passage.

No. 1 is the frame that guides the piles strait.

No. 2, pices of timber across over the back of the arch, to keep the foot of the piles firm when we move the Casseon.

No. 4, these piles will have a half-circle cut oute of each, of about 3 In Diamr, to ram down oakem to keep the piles tight in the joints.

This plan will verry much shorten the tunnel as it is not so deep, therefore will com to day much sooner on each side [of] the river, and all the timber that must have been to support the roofe will now be saved. Your opponion on this plan wod verry much oblige Your verry

Humble sert,

Rd TREVITHICK.

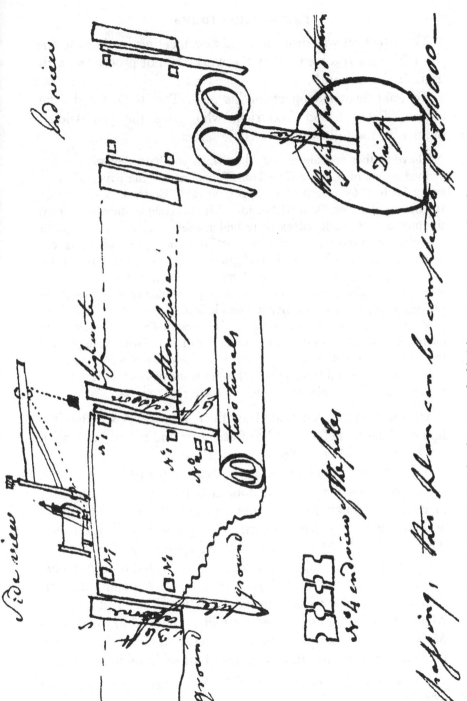

Fig. 18. Trevithick's plan for a tunnel of brick, in sections, under the Thames, 1808

The situation was discussed and at a meeting on August 4th the Directors resolved "that the works be not proceeded with until further orders".

To gain further support for his plan, Trevithick sought the advice of his friend Goodrich, who gave his professional opinion in the following ostensible letter:[1]

I have further considered the Plan which you have devised and explained to me of fixing a Cast Iron Tunnel across the River Thames under its bed, by means of an upper movable Caissoon and a lower surrounding Coffer Dam withinside of it, to exclude the water from the portions of work successively under execution, and to afford an opportunity of excavating the ground from above, and of fixing the Tunnel in this trench open to daylight, so as to be, when covered in, sufficiently below the bed of the River to be protected from injury— and it is my opinion that this Plan may be proceeded upon, under good management, with a certainty of completion.

With respect to the enlarging of the present Drift Way so as to form a Tunnel of it twelve feet in diameter or only large enough for a foot passage I am of opinion after what you have explained to me of the nature of the ground to be passed thro' in some places that it is nearly impracticable, and quite so under any moderate expense.

It will be observed that Trevithick now proposes that the lining of the tunnel should be of cast-iron and not of brick as in the letter to Giddy—quite modern practice.

The Directors held a meeting on November 16th to decide as to Trevithick's plans, when doubtless he adduced Goodrich's favourable opinion in their support. The Directors instead of adopting them resolved to invite the opinions on it of General Twiss, Lt.-Col. Mudge, Lt.-Col. Shrapnell and Sir Thomas Hyde Page, Royal Engineer officers it is true, but none of whom had any experience to justify an opinion. We assume that the Directors, on their advice, turned down Trevithick's plan, for on March 30th, 1809, they issued a public invitation[2] to "ingenious Men of every description to a consideration of the best means of completing so useful, and so novel an undertaking", offered

[1] Goodrich Papers, Science Museum, South Kensington. Press copy dated 1808, Sept. 5th.

[2] Copy among the Goodrich Papers, Science Museum, South Kensington.

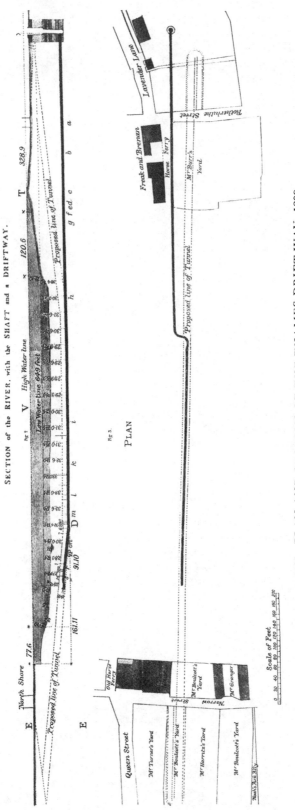

SECTION of the RIVER, with the SHAFT and a DRIFTWAY.

PLAN

PLATE XII. PLAN AND SECTION OF THE THAMES DRIFT-WAY, 1809
From a lithograph issued by the Directors

a premium of £200 for a plan that "shall be adopted and acted upon", and a further sum of £300 if it be executed. For the information of the competitors the Directors had prepared a plan[1] of the work (see Plate XII) with descriptions and particulars of the strata passed through. These documents throw much light on what had actually been accomplished.

No less than fifty-four plans were submitted by the first of June in response to this notice and were referred to Dr Hutton, the celebrated mathematician, and Mr William Jessop, the well-known civil engineer, but they "agreed in opinion that it was impracticable to make under the Thames a tunnel of useful size by an underground excavation".[2] This was final if anything could be.

Tunnelling under the Thames had to wait ten years till Brunel had invented the tunnelling shield, and a further decade till it had been perfected.[3] This is not to say that, even with the means available in 1808, Trevithick might not have been successful with his proposed tunnel in sections—a method that was used in 1906–9 across the Detroit river, Michigan, and again quite recently in San Francisco Harbour.

Let us admit that the drift-way was a failure, yet it was a glorious failure. In six months Trevithick had driven over 1000 ft. of heading out of a total of 1200 ft.—a labour of Hercules. Our admiration is stirred by his physical courage and by the engineering skill he displayed, which justifies our opinion that there was no branch of the profession of civil engineering that he did not adorn.

Now that active work on the Thames drift-way was brought to a dead stop, Trevithick, finding time on his hands, was naturally open to a proposal which gave some hope of resuscitating the application of his engine to transport, thereby realizing the con-

[1] Copies among the Goodrich Papers, Science Museum, South Kensington.

[2] *The Thames Tunnel—An exposition of facts submitted to the King by Mr Brunel*, 1833, 4to.

[3] Beamish, *Memoir of the life of Sir Marc Isambard Brunel*, 1862, pp. 202–225.

viction to which he had given expression in his letter to Sir John Sinclair, quoted above (p. 69). The origin of this fresh proposal may best be inferred from the notice which appeared in *The Times*, July 8th, 1808:

We are credibly informed that there is a Steam Engine now preparing to run against any mare, horse, or gelding that may be produced at the next October Meeting at Newmarket; the wagers at present are stated to be 10,000*l*; the engine is the favourite. The extraordinary effects of mechanical powers is [*sic*] already known to the world; but the novelty, singularity and powerful application against time and speed has created admiration in the minds of every scientific man.—TREVITHICK, the proprietor and patentee of this engine, has been applied to by several distinguished personages to exhibit this engine to the public, prior to its being sent to Newmarket; we have not heard this gentleman's determination yet; its greatest speed will be 20 miles in one hour, and its slowest rate will never be less than 15 miles.[1]

In the same newspaper on July 12th appeared the following paragraph:

The Steam Engine mentioned in our Paper of Thursday last, to run at Newmarket next October Meeting, will be completed this week, and exhibited to the public in the fields adjoining the Bedford Nursery, near Tottenham-court-road of which due notice will be given in this Paper.

Another interesting notice which embodied a little further information appeared in the *Observer*, Sunday, July 17th:

The most astonishing machine ever invented is a steam engine with four wheels so constructed that she will with ease and without any other aid, gallop from 15 to 20 miles an hour on any circle. She weighs 8 tons and is matched for the next Newmarket meeting against three horses to run 24 hours starting the same time. She is now in training on Lady Southampton's estate adjoining the New-road near Bedford Nursery, St. Pancras. We understand that she will be exposed for public inspection on Tuesday next.

[1] We are indebted to the late Mr R. B. Prosser for making this search; his results were embodied in an article in *The Engineer*, Dec. 18th, 1903, p. 589, *q.v.*

The promised advertisement duly appeared in *The Times* of July 19th thus:

RACING STEAM-ENGINE.—This surprising Engine will commence to exhibit her power of speed to the Public, THIS DAY, at 11 o'clock, and will continue her experiments only a few days. Tickets of admission, 5s. each, may be had at the Bar of all the coffee-houses in London; and at the Orange Tree, New-road, St. Pancras.

However, the engine did not start as advertised because of a hitch due to yielding of the earth under the track. The following advertisement was accordingly inserted in *The Times*, July 20th, to explain the delay:

RACING STEAM-ENGINE.—The Public are respectfully informed, that the exhibition of this machine, which was to have taken place this day is unavoidably POSTPONED till Monday next, the ground under the Railway, on which it was to run, being too soft and spongy, requiring additional support of timber.

This is corroborated by a letter from Trevithick to Giddy:[1]

I have yours of the 24th, and intend putting the inscription on the engine which you sent me. Abt 4 or 5 days since I tryd the engine which workd exceeding well, but the ground was very soft, and the engine, abt 8 tons, which sunk the timber under the rails and broake a great number of them, since which I have taken up the whole of the timber and Iron, and have laid Balk of from 12 to 14 In Square, down in the ground, and have nearley all the road laid again, which now appair to be very firm, as we prove every part as we lay [it] down, by running the engine over it by hand. I hope that it will all be compleat by the end of this week.

The "inscription" referred to is probably the name "Catch me who can" which was given to the engine by the sister of Davies Giddy, Mrs Guilmard, who made the following memorandum: "My ride with Trevithick in the year 1808 in an open carriage propelled by the steam engine of which the enclosed is a print took place on a waste piece now Torrington Square".

The "print" to which she refers is that of a card which apparently Trevithick had had engraved for the occasion (see

[1] *Enys Papers. Life*, i, 273. Letter dated Thames Archway, Rotherhithe, 1808, July 28th.

Fig. 14), no doubt a card of admission to the enclosure. Fortunately more than one copy has survived[1] and indeed it is the only evidence we possess of the appearance of the engine. The title is "TREVITHICKS PORTABLE STEAM ENGINE 'Catch me who can'"; the sub-title "Mechanical Power Subduing Animal Speed" quite confirms the newspaper accounts of the origin of the experiments.

Fig. 14. Admission card to Trevithick's railway, 1808. From the *Enys Papers*

Mrs Guilmard's memorandum can hardly have been written before the late 'thirties when the square mentioned was completed (only one side called Torrington Street was in existence in 1808), hence the evidence is not as valuable as if it were contemporary.

The other print[2] which is preserved in the *Enys Papers*,[3] has the following written on it:

[1] In the *Enys Papers*.   [2] *Life*, I, 192.
[3] Reproduced by Miss Edith K. Harper, in her valuable monograph *A Cornish Giant*, 1913. The card measures $3\frac{1}{2}$ in. × $2\frac{1}{2}$ in.

1804. I'rode on a steam carriage made by Trevithick at Merthyr Tidville in South Wales. D. Gilbert.

A bet was made that this machine would go further in 24 hours than a race horse... and a circular enclosure was made near Gower St. for this to move in, about 1814.

Obviously again this was written from recollection subsequent to 1816 when Giddy changed his name to Gilbert.

The wager must have given rise to much gossip and we have a contemporary confirmation of this in the newspaper reports of the trial Chapman v. Thames Archway Co. in which, on July 25th, Trevithick appeared for the defence. He is spoken of as "the inventor of the steam engine which is to run at Newmarket". Another contemporary reference is to "the new *trotting match* in which one of the competitors is to be a steam engine".

How soon the engine was ready, and the public admitted to see and ride upon it, we do not know. Presumably it was some time during the two months that elapsed before we hear of it again when the bet was to be decided. The announcement appeared in the *Observer* of September 18th, 1808, and runs thus:

EXTRAORDINARY WAGER.—It has been some time announced, that the NEW MACHINE for travelling without horses, being impelled entirely by STEAM, was matched to run twenty-four hours against any horse in the kingdom. This bet, so novel in the sporting world, will be decided on Wednesday and Thursday next. The machine is to start at two o'clock on Wednesday, on its ground in the fields, near Russel-square, to demonstrate the extent of its speed and continuance. It is calculated that the machine, though weighing eight tuns, will travel 240 miles, at least, within the time limited.—Very large sums are depending on the issue.

It must have been this or a similar paragraph that caught the eye of Samuel Homfray and caused him to write to Messrs Haynes and Douglas, in London:[1]

I wish likewise you to say if there is anything in the report of the Racing Machine being carried into effect and if so at what time, and the

[1] *Life*, I, 235. Letter dated Penydarren Place, 1808, Sept. 24th.

particulars of the bet &c., as if it is to take place, I shall be very much inclined to see it.

Obviously this was just the kind of event in which a sporting character, such as Homfray was, would have revelled and upon which he would have liked to have had money himself. He does not seem to have given a thought to the technical aspect of it; our only comment is, "so English you know". However, at the time of writing the trials were over, presumably, for we do not find any further notices in the press.

A few crumbs of information were given by John U. Rastrick, who had worked as an engineer under Trevithick on the arch-way, when giving evidence on the Liverpool and Manchester Railroad Bill, in 1825; he said:

About ten or twelve years ago I made one [i.e. a locomotive] for Mr Trevithick, the person who had the original patent for making it; this was exhibited in London; I did not see that myself. A circular railroad was laid down and it was stated that this engine was to run against a horse and that whichever went a sufficient number of miles was to win.

A better account, although unfortunately forty years after the event, is that of John Isaac Hawkins (1771–1855); it is worth quoting because he was an engineer and a reliable observer:[1]

Mr Trevithick's New Road Experiments in 1808.

Sir,

Observing that it is stated in your last number [No. 1232, dated the 20th instant, page 269], under the head of "Twenty-one Years' Retrospect of the Railway System", that "the greatest speed of Trevithick's engine was five miles an hour", I think it due to the memory of that extraordinary man to declare that about the year 1808 he laid down a circular railway in a field adjoining the New Road, near or at the spot now forming the southern half of Euston Square; that he placed a locomotive engine, weighing about 10 tons, on that railway, on which I rode, with my watch in hand, at the rate of 12 miles an hour; that Mr Trevithick then gave his opinion that it would go 20 miles an hour, or more, on a straight railway; that the engine was exhibited at one shilling admittance, including a ride for the few who were not too timid; that it ran for some weeks, when a rail broke and

[1] *Mech. Mag.* XLVI, 1847, p. 308.

occasioned the engine to fly off in a tangent and overturn, the ground being very soft at the time.

Mr Trevithick having expended all his means in erecting the works and inclosure, and the shillings not having come in fast enough to pay current expenses, the engine was not set again on the rail.

I am, Sir, your obedient servant,

JOHN ISAAC HAWKINS,
Civil Engineer.

The exact spot where the trials took place has been the subject of controversy.[1] All the accounts agree that the site was within the area bounded by the New (now Euston) Road on the north, Tottenham Court Road on the west, the Bedford Nurseries (roughly where Endsleigh Gardens are now) on the east and Francis Street on the south, for that street is roughly the boundary of the Southampton Estate.[2]

A glance at a map of London of about the date in question (see Fig. 15) shows that there was a considerable area, then unbuilt upon, where University College now stands. It was on this site, the authors believe, that the trials took place. It is intended that a Memorial Tablet to record the event shall be erected there by the Trevithick Centenary Commemoration Committee.

Francis Trevithick gives a drawing of the scene of the trials, by W. J. Welch, and a drawing with a strong family resemblance to it, attributed to Rowlandson, was picked up in Charing Cross Road some years ago.[3] The result of an examination of this drawing by Mr Martin Hardie, of the Victoria and Albert Museum, is that it is neither of the technique of Rowlandson (1756–1827) nor is the paper on which it is drawn of his period. Two other drawings of equally close resemblance have come to light this year: one is at Merthyr Tydvil Library and the

---

[1] Cf. C. R. King in *The Locomotive*, June 14th, 1930, p. 200.

[2] As will be expected, no record exists at the office of the Bedford Estate, which is just south of the Southampton Estate, of any trials in 1808; indeed the suggestion for a "roundabout", for that was what the show would have seemed, on such a "select" estate would, if made, have been rejected with disdain.

[3] Now in the possession of Capt. R. E. Trevithick.

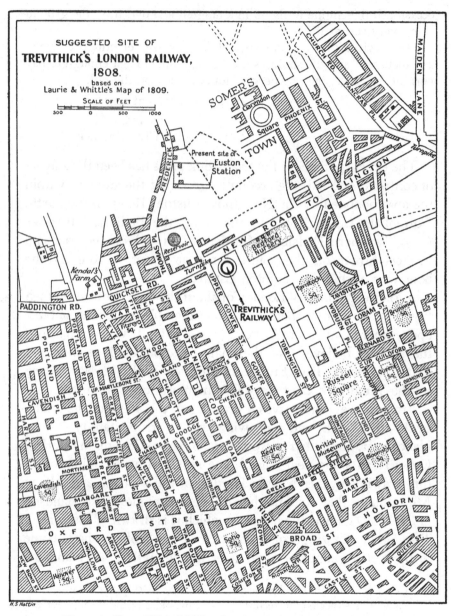

Fig. 15. Site of Trevithick's London railway, 1808, on a map
of the period

other in private possession. These the authors have not had examined, but it may be sufficient to say that the former of the two shows what appears to be a hobby horse (i.e. *c.* 1818), and a disc railway signal (i.e. *c.* 1842). The authors do not attach any importance to these pictures.

As regards the locomotive, if we accept the engraving (Fig. 14) as a representation of it, and the authors do accept it, we can say that it was remarkably simple in design; in fact it was the small dredger engine (Fig. 11) placed on wheels. These wheels were neither coupled nor geared; they were smooth and ran on smooth rails which were probably of plate section, such as were used on the tramways of that date. The rails might even have been brought from the drift-way.

The reason for discontinuing the show, for such it was, is no doubt that given in Hawkins's letter—lack of patronage, shortness of funds and probably a conviction on Trevithick's part that he had established all he set out to do. At any rate it was his last gamble with the locomotive.

It will be convenient to interpose here, in order to give some clue to Trevithick's subsequent actions, what we know about the property in the patent. At a very early stage, if not from the beginning, William West was admitted as a partner and the shares were: Trevithick two-fifths, Vivian two-fifths and West one-fifth. Homfray was anxious to come in in 1803 and eventually did so in 1804 or 1805 by buying half the patent. Vivian was in need of money and wished to sell out his share altogether. He was in treaty with Sir William Curtis in 1805 when the sum of £4000 was mentioned as its value, but this seems to have led to no result and he sold his share to a Mr Bill, who would appear to have been merely a nominee of Homfray; the latter, therefore, held a controlling interest. In 1807 the position was: Homfray one-half, Trevithick one-fifth, Bill one-fifth and West one-tenth. Trevithick now wished to dispose of his share to Messrs Haynes and Douglas, cotton merchants, of Tottenham Court Road. What his motive was, except want of cash, and why cotton merchants should want to buy his share, is a mystery. At any rate Homfray, in a letter to Haynes and Douglas dated September 24th, 1808,[1] refers to the firm as "being interested in the concern", so that we conclude that by that date Trevithick had parted with his last holding in the patent when it had still eight years to run. West it would seem retained his tenth to the very end.

Trevithick's horizon had now been greatly widened. He had matched himself against many leading engineers and come into contact with prominent business men and with public bodies. He had brought his wife and family up to London and was living in Limehouse, then on the fringe of London, and not yet covered with houses, canals, docks, factories and warehouses. Is it to be wondered at that some person, seeing a man of Trevithick's stamp, teeming with ideas, should have thought there was an opportunity to exploit them? Such a person with such an outlook Robert Dickinson seems to have been. He is stated to have been a West India merchant, but judging by his record at the Patent Office he should be called, very justifiably, a professional inventor, for between 1801 and 1826 he took out no less than twenty-three patents of the most varied description.

It seems likely that Dickinson thought that the best way of securing his own interest and of turning Trevithick's ideas into money was to induce him to become a joint patentee, possibly supplying the requisite cash—a patent cost about £100 in those days—the necessary assistance in connection with the specification and carrying through the legal formalities before receiving the Great Seal, all of which would have been, we are sure, utterly repugnant to Trevithick's disposition.

Together, Trevithick, describing himself as of Rotherhithe, engineer, and Dickinson, describing himself as of Great Queen Street, esquire, took out two patents in 1808 and one in the following year.

The first, No. 3148, July 5th, was for what they called a "nautical labourer", i.e. a floating crane and tug for dock and

---

[1] *Life*, I, 235, where the correspondence (preserved in the Science Museum, South Kensington) is quoted. The sequence of events is obscure but the résumé given above is the outcome of a close study of the documents.

harbour duties at a time when nothing more effective than a capstan or hand winch was to be had. To quote from the patent specification:

The novelty of our invention is simply this: employing such a vessel as we have described, furnished with a steam engine as a moving power, and with proper apparatus to enable us to employ the said vessel and its contents as a labourer to assist in towing of vessels in the manner before described, and in loading and unloading them, in place of using methods hitherto in use.

For towing purposes the engine of the "labourer" drove a single paddle-wheel amidships in an air-tight chamber with an air pump to keep down the water level inside. The idea of such a paddle-wheel has been a favourite one with inventors. We do not know whether such a "labourer" was actually constructed, but Francis Trevithick says:[1]

Several experiments were made in the Thames, with every prospect of an immediate and extensive use of the nautical labourer, when the Society of Coal Whippers protested against the encouragement of such a rival, declined to work with it, and threatened to drown its inventor. Trevithick was guarded by policemen, two of them keeping watch at his house; but he succeeded in proving that his idea was economical and practical, though it was impossible to stem the strong current of prejudices of the daily labourer, and the nautical labourer retired from the contest.

We are strengthened in the opinion that Dickinson was just exploiting Trevithick, because the latter had proposed the very same idea to Giddy two years earlier:[2]

A Genelman have orderd an engine for Driveing a ship. Its a 12½-In Cylinder. I am at a loss how to construct the apperatus for this purpose; therefore am under the necessity of troubleing you for your advice on the subject. The plan I have is as under, unless you condem it, or sugest a better plan. I propose to put a horizontol engine below the deck, and to put a Wheel of 14 ft Diam in the hold. This Wheel is to work in an Iron case, air-tight; the axle to work in a stuffing box, and a pump to force air into this case to keep down the water from flooding the wheel, so that onely the flotes on the extremity of the

---

[1] *Life*, I, 334.

[2] *Enys Papers. Life*, I, 327. Letter dated Blackwall, 1806, July 23rd.

wheel shall be in water, and then onely extend abt 15 Inches below
the keel of the ship. The cutter is abt 100 tons burthon. I wish to know
the size of the flotes on the wheel, and the volosity you wod propose
the[y] shod be drove. I think the power of the engine is equal to
400,000 one foot high in a minute, from which you will be able to
judge what size flotes, and volosity wod be best.

Below you have the proposed plan. The air thats forced in to this
wrought Iron case will allways keep the water down in this case to the
level of the bottom of the ship, and a space will be left on each side the
wheel, so that the air will never be dissplaced by the working of the
wheel.

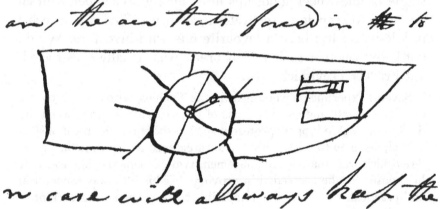

Fig. 16. Trevithick's steamboat with paddle wheel amidships, 1806

Furthermore, there is evidence that for unloading ships
Trevithick had schemed a perfectly satisfactory arrangement as
early as 1804. The description of the apparatus and the reason of
its being set aside is given in the following letters:[1]

There is an engine order'd for the West India docks to travel itself
from ship to ship to unload and to take up the goods to the upper
floors of the store houses at the crane and in case of fire to force water
on the store houses. The fire is to be keep'd constaintley burning in
the engine so as to be ready at all times.

The engine that was made for the London docks for dischargeing
the west Indiamen is put to work in a manfactorey. The[y] wod not
permit fire within the walls of the Dock; there is [an] act of parleyment
to that effect. There is a small engine makeing at Staffordshire, for the

---

[1] *Enys Papers.* Letter to Giddy, dated Penydarran, 1804, Feb. 22nd.

London Coal ships to carry with them for unlodeing. The boiler is 2 ft 6 In Diam and 3 feet long with the cylinder horizontal on the outeside [of] the boiler, 18 In stroake, Cylinder 4 inches [diameter]. I think it will be equal to 6 men. The engine will always be going one way and the man that stands at the hatch way will have a string to throw out a catch which will let the barrall run back with the empty baskete.[1]

To put the matter in a nutshell, Trevithick's "nautical labourer" was shut out in 1805 by statutory limitations which shipowners would not bestir themselves to remove, and in 1808, consequently, it had no chance of adoption.

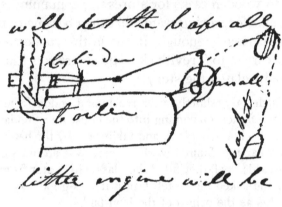

Fig. 17. Trevithick's high-pressure engine for discharging cargo, 1804

Trevithick and Dickinson's second joint specification is dated October 31st, 1808, and was for stowing cargo in iron tanks instead of the then universal wooden casks. The patentees published a pamphlet descriptive of their invention dedicated "To His Majesty's Ministers, the Rt. Honourable the Lords of the Admiralty and...the Shipping Interests of Great Britain", and dated from 58, Great Queen Street, Lincoln's Inn Fields (i.e. Dickinson's address), February 10th, 1809. As an instance of the enormous saving effected by their patent iron tanks as compared with wooden casks the inventors state: "A ship engaged in the whale fishery belonging to Messrs Burnet & Co. lately returned from a successful voyage". Her burden was 208 tons

[1] *Enys Papers. Life*, i, 324. Letter dated Soho Foundry, Manchester, 1805, Jan. 10th.

and yet with her hold completely full she held only 140 tons of oil, "being fully one-third less than what she might have stowed in differently formed packages of iron". The patentees go on to state that they have engaged to furnish this ship with iron tanks. They give a comparative estimate showing that by the increased stowage the profit would amount to an additional £9000 on a single voyage. A letter from Sir Humphry Davy—one Cornishman always backs up another—is appended, dated Royal Institution, February 18th, 1809, stating his opinion that water cannot be injured by storage in air-tight iron tanks and that they are preferable to wooden casks for stores not containing salt.

It is as tanks for storing water that they became of the widest service and, curiously enough, it was in this connection that the idea first occurred to Trevithick. His eldest son Richard, then ten years old, related the story:[1]

When a little boy, shortly after reaching London from Cornwall, about 1808, my father, on coming into the house on a Sunday morning, desired me to fetch a wineglass, and taking me by the hand, walked to the old yard near the Tunnel works. There was an old steam-engine boiler in the yard; my father filled the glass with water from the boiler gauge-cock, and asked me to tell him if it was good water. We used to speak of this as the origin of the iron tanks.

The invention has been an incalculable boon to both the merchant and the naval services, not merely for storing water but as ballast tanks when leaving a port light, to obviate the practice we have mentioned above of using actual ballast from the river or shore.

The patentees entered into partnership and set up a manufactory for tanks at 72, Fore Street (now renamed Narrow Street, see map, p. 91), Limehouse, whither Trevithick had removed on the closing down of the Thames Archway. John Steel, a Newcastle man, who had worked there for Trevithick on the locomotive, was foreman, assisted by Samuel Hambly, a cousin of Mrs Trevithick, and Samuel Rowe. The two latter were Cornishmen and had worked on all the locomotive engines. There appears to have been disagreements between Trevithick

[1] *Life*, i, 287.

and Dickinson over technical as well as money matters and the manufactory seems to have been carried on at a loss, and well it may, for neither of them seemingly attended to the business, but left it to the men.

An extension of the idea of his tanks, emanating from Trevithick's fertile brain, was to use them for raising wrecks by supplying buoyancy, on the method now so widely known and practised. In January and February, 1810, Trevithick and Dickinson undertook to raise a wreck at Margate.[1] Mrs Trevithick stated that the attempt was successful, and the wreck "was being floated into shallower water when a dispute arose about the payment. Her husband wished an immediate 'yes' or 'no' to his bargain. The shipowners wished to defer the answer until the ship was safe. In a moment the tanks were cast loose and the vessel was again a sunken wreck". If this is a true story, it is only another instance of unbusinesslike proceeding and hot-headed impetuosity on the part of Trevithick.

This novel application of the iron tanks was sufficiently hopeful to induce Trevithick and Dickinson to petition for Letters Patent on March 23rd, 1810, for "Certain Inventions or new Applications of known Powers to propel Ships and other Vessels employed in Navigating the Seas or in Inland Navigation to aid the recovery of Shipwrecks, promote the health and comfort of the Mariners and other useful purposes", but nothing came of the petition possibly because Trevithick's illness supervened.[2] The iron tank was a capital invention which ought to have been a little gold mine; but it is stated that Trevithick made no profit out of it. Henry Maudslay took over the patent, under circumstances to be detailed later, and out of it got substantial returns, as a side-line in his steam-engine manufactory at Lambeth.

The third joint patent of Trevithick and Dickinson (April

[1] *Life*, I, 298.

[2] The sign manual warrant of the King to the Law Officers for the preparation of the Patent Bull is entered in the Home Office Warrants Book. The warrant itself is among the *Enys Papers*; how it could have got into private possession we can only guess.

29th, 1809, No. 3231) was principally for the use of iron in shipbuilding. The clauses are (1) a floating graving dock, (2) ships of iron for ocean service, (3) telescopic masts and yards of iron, (4) seasoning and bending timber by heated air, (5) diagonal framing for ships, (6) buoys made of iron, (7) crane attachments to the "nautical labourer" previously mentioned, (8) propulsion of a vessel by a "rowing trunk" or piston displacer, (9) steam galley and fresh-water condenser.

Most of these were extraordinarily fertile ideas, albeit not all of them were new; any one of them needed years of labour to bring to perfection or carry into practice. As before, the patentees published a pamphlet[1] dated February 10th, 1809, dedicated as in the previous one, describing the inventions and giving a statement of the advantages; with the exception of the iron buoys, which were made at Limehouse, nothing seems to have been done with the patent although Trevithick seems to have got out designs under it for an iron sailing ship.[2]

In the midst of these employments, in May, 1810, he was stricken by a serious illness—the only serious one he ever had— which incapacitated him for more than six months. How it was brought on we do not know, but it proved to be a turning-point in his career.

His son says:[3]

Typhus and gastric fever during many weeks reduced him to a state of physical helplessness, followed by the loss of intellect, and brain fever, and the patient, before so weak, required the care and strength of keepers.

In this emergency Dickinson brought his medical man to assist; Dr Walford, known to the family, disagreed with the new comer. A third medical man was called in. Anonymous letters were sent to Mrs Trevithick on the probable result of the injurious medical treatment.

Francis Trevithick goes on to describe the frantic state to which Mrs Trevithick was reduced by anxiety for her husband's life. Fortunately, however, his constitution enabled him to battle with the disease and we soon hear of him as having got

[1] The writers have not been able to trace a single copy of this pamphlet.
[2] *Life*, I, 316, 318.   [3] *Life*, I, 338.

over the crisis. Mrs Trevithick's brother, Henry Harvey, whose kindness can hardly be praised too highly, wrote as follows:[1]

MY DEAR JANE,

Our hearing by every other post from Mr Blewett, of Trevithick's rapid recovery, and also by Dr Rosewarne last Saturday, that the fever had quite left him, gave us great satisfaction; but we are much concerned for your situation.

In your letter of the 25th ult. you seemed to be much alarmed from Trevithick's weakness, but I think you cannot expect otherwise than that he will be very weak for some time, after so dreadful an attack. Do not be alarmed, I hope he will do very well; you must not say anything to him about his business, that is likely to hurry his mind, until he gets better. If Mr Dickinson receives money, he must be accountable for it. I beg that you will not hesitate asking Mr Blewett for what you want. It gives us great happiness to hear that you enjoy health in this great trial. If you think it necessary and you wish it, I will come to you, but I sincerely hope that your next will bring a more favourable account; I know Mr Blewett will be very happy to do anything for you in his power, and I wish you would ask his advice in any business that you think is not proper for Trevithick to be told of until he gets better.

I hope you find Mr Steel honest: in that case it is not in Mr Dickinson's power to cheat you.

Do let me know how Trevithick's affairs stand, and what his prospects are. If he is not likely to do well where you are, do you think he would consent to return to Cornwall—if not to settle, for a little while? His native air might be a means of getting him about. Both my sisters join with me in love to you and family.

We can imagine with what joy Mrs Trevithick would hail the suggestion to leave dirty and smoky Limehouse; indeed she must have longed many a time for home. We can see her deciding at once that her husband should return, but convalescence proved to be slow, for the attack had been most severe.

Francis Trevithick's account of the home-going, which incidentally throws a sidelight on the state of European affairs at the time, is as follows:[2]

In the early part of September, 1810, being still too weak to move

[1] *Life*, I, 340. Letter dated Hayle Foundry, 1810, June 1st.
[2] *Life*, I, 339.

hand or foot, he was carried on board a small trading vessel, called the "Falmouth Packet", his eldest boy, about ten years old, keeping him company.

It was war time, and the "Falmouth Packet" with other vessels sailed from the Downs, under convoy of a gun-brig. After three days they anchored off Dover. Trevithick went on shore and enjoyed the first short walk since the commencement of his illness, four or five months before. On getting under way again they were chased by a French ship of war. The "Falmouth Packet" knew how to sail and when to hug the shore, so she showed her heels to the enemy; and in six days after leaving London landed him at Falmouth, about sixteen miles from his Cornish residence at Penponds. Taking his boy by the hand, they walked to his home, from whence two months before, when too ill to be informed of his loss, his mother had been carried to her grave from the house of his childhood.

More than three months elapsed before he recovered his accustomed health and vigour, helped thereto by his fine physique, the care of his wife and the tonic air of his native Cornwall.

But the sword of Damocles was hanging over him. Early in the year before his illness, his son says that "he had gone [to Cornwall] to raise money by mortgage or sale of mine shares or property". During his illness, his affairs in London went from bad to worse, and the crash came on February 5th, 1811, when the partners were declared bankrupt.

The notice appeared in the *Gazette*[1] in the stereotyped form:

Whereas a Commission of Bankrupt is awarded and issued forth against Richard Trevithick and Robert Dickinson, late of Fore Street, Limehouse, in the county of Middlesex, Dealers in Iron Tanks, Dealers, Chapmen and Co-partners and they being declared Bankrupts are hereby required to surrender themselves to the Commissioners in the said Commission named or the major Part of them on the 16th and 23d of February instant, and on the 23d of March next, at Twelve of the Clock at Noon, on each of the said Days, at Guildhall London, and make a full Discovery and Disclosure of their Estate and Effects....

The solicitors were Wadeson, Barlow and Grosvenor, of Austin Friars.

Trevithick must have come up to London to attend at the

---

[1] *London Gazette*, Feb. 9th, 1811, p. 245.

PLATE XIII. TREVITHICK'S HOME AT PENPONDS, AS IT APPEARED IN 1870

Courtesy of the Science Museum, South Kensington

Guildhall for his examination on February 6th, February 23rd, and March 23rd, and have submitted a statement of his affairs. It is this presumably that his son refers to when he says:[1]

An official document of receipt and expenditure for the five years preceding his bankruptcy shows a loss of more than 4,000*l.* by his labours; and that he had sold or mortgaged his property and borrowed from his friends.

We take this to mean that the debts of the partnership amounted to £4000 and that the assets were practically nil, but as a matter of fact the composition paid, as will be seen below, was about 16*s.* in the pound. For the time being, however, the effect was disastrous. His son continues the sad story:[2]

Everything belonging to him was seized for debt, and he was obliged to retire to a sponging-house in a street of refuge for debtors, a halfway house between freedom and imprisonment. All this was too much for the strongest man.

In writing thus Francis was assuming that his father's bankruptcy occurred before his illness and was a contributory cause, whereas dates show the reverse to be true. We trust that Dickinson was laid by the heels, too, but it is more than likely that he was clever enough to escape the clutches of the law. Trevithick seems to have felt keenly that he had been injured; at any rate he was under a cloud for nearly a whole year, during which we hear practically nothing of him.

Trevithick himself makes a few allusions to this unfortunate event:[3]

As I have been a bankrupt, perhaps you may scruple on that account [i.e. to lend money], but that business is finally settled, and I have my certificate; and indeed I never was in debt to any person: not one shilling of debt was proved against me under the commission, nothing more than the private debts of my swindling partner.

The date of discharge does not agree with that given below, by the way, but that is a small matter.

---

[1] *Life,* I, 341. [2] *Life,* I, 338.
[3] *Life,* II, 172. Letter to Sir C. Hawkins, dated Camborne, 1812, March 10th.

In a letter[1] to Rastrick a year later Trevithick makes some further comments *apropos* of money which a certain Mr Shuttleworth claimed to be due to him:

...the Bankrup^{cy} was brought on the partnership concern and not for my private debts....Dickinson entered in the Deeds with me that he was to pay every demand...I...loosed all I had in the concern, there will be about 16s. paid in the pound. The reason why I have not had a certificate is because Dickinson stole the papers from the attorney office but there never was any objection to their granting it and the Attorney said he would immed^{ly} wait on the Lord Chancellor for me because it cannot be done in the usual way.

This is somewhat cryptic, but it is explained by an order dated August 7th, 1813, of the Lord Chancellor, which recites the circumstances:[2]

That the last Exam^n of the sd Bankts was adjourned from time to time and was finally taken on the 4th day of May 1811 soon after which time the sd depositions and proceedgs under the sd Comm^n were lost or mislaid and have not since been found.

The gist of it was:

I do order that the Detrs be at liberty to call a meeting of the Commrs...in order that the Cred^{rs} who have already proved their Debts...be at liberty to reprove the same.

No doubt this was done, for the bankrupts received their discharge on January 1st, 1814,[3] and a meeting was called on April 18th, 1815, "to make a Further Dividend of the Joint Estate and Effects of the said Bankrupts".[4] Presumably this was the final dividend and the estate was then wound up.

It is pretty clear that some shady work had gone on and we are disposed to look upon Trevithick as an innocent and unsuspecting victim; in other words this was a case of the kind, more common then than in these days of limited liability, where a man became involved financially by an unscrupulous partner.

Here we leave Trevithick for the moment to recover from his misfortunes and to begin life again among his kith and kin.

[1] *Draft Letter Book.* Letter dated Camborne, 1813, March 12th.
[2] Record Office. Bankruptcy Order Book, 1811–13, No. 122.
[3] *London Gazette*, Dec. 11th, 1813, p. 2505.
[4] *London Gazette*, March 25th, 1815, p. 273.

CHAPTER IV

INVENTION AT FLOOD-TIDE

Wheal Prosper, first Cornish engine—Expansive working—Thrashing engines—Steam cultivation—Sugar mills for West Indies—Plymouth breakwater—Tour with Rastrick in Cornwall and Devon—Plunger-pole engines at Wheal Prosper and at Herland—Cornish boiler—Compounding engines—Rivalry with Arthur Woolf—Recoil engine—Screw propeller—Characteristic conduct.

TREVITHICK'S recovery from his serious illness must have been retarded, we fear, by the knowledge that bankruptcy was hanging over him; the blow fell, as we have said in the last chapter, in February, 1811. He had now to start life over again, for everything of material value went in the wreck. His brains, his energy and his devotion to the high-pressure engine and boiler remained. He must have been quite recovered by the new year of 1811, for at that time we learn of him writing to Giddy on the subject[1] of utilizing the greatest possible amount of the heat of fuel by passing the combustion products through the water in the boiler and utilizing the mixed gases and steam so produced in the cylinder. The idea seems to have been suggested to him by his proposal to smelt copper in a blast furnace. Trevithick says: "Prehaps its like maney other wild fanceys that flyes through the brain; but did not like to let it go onnoticed withoute first getting your opponion, which I hope you will excuse me for so often troubleing you".

From Giddy's reply we learn[2] that Trevithick had suggested to Giddy the idea of thus smelting copper ten years earlier. Giddy's conclusion about the direct-heated boiler is: "This plan will certainly not do. Write to me by all means whenever anything strikes you and you may always depend on having my best advice". Naturally the matter went no further. Not that the

[1] *Enys Papers. Life*, ii, 6. Letter to Giddy, Camborne, dated 1811, Jan. 13th.

[2] *Life*, ii, 10. Letter dated London, 1811, Jan. 20th.

idea is chimerical, for it has since engaged the attention of inventors, and very large sums have been spent on the solution of the problem, hitherto without complete success, owing to the enormous practical difficulties to be surmounted.

Casting about him as to what he should take up, we imagine that he would revert to the employment he had before leaving Cornwall, in other words what we should to-day call a consulting practice.

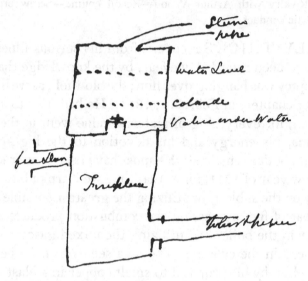

Fig. 18. Trevithick's boiler with direct contact of fuel
gases and water, 1811

In the letter to Giddy, just quoted, Trevithick alludes to a rough survey which he had made, apparently for Sir Christopher Hawkins, Bart., M.P., of the harbour of St Ives for a breakwater, but the scheme was not proceeded with.[1] Apparently he resumed his efforts to get mine managers to apply higher pressures to their large pumping engines. The course of events can be pieced together, only somewhat disjointedly, from the recollections of mine captains and engine-men, and from passages in Trevithick's correspondence. His first important job was with Dolcoath Mine. He renewed the proposals, broken off in

[1] *Life*, II, 16.

1806,[1] to apply his improved boiler to the existing engines. His proposals were accepted and the boilers were made at the mine under his directions. They replaced the existing waggon boilers, which were probably worn out in any case. The only evidence of this comes from the recollections of Captain Nicholas Vivian in 1858, and is as follows:[2]

About 1812 Captain Trevithick threw out the Boulton and Watt waggon boilers at Dolcoath and put in his own, known as Trevithick's boiler. They were about 30 feet long, 6 feet in diameter, with a tube about 3 feet 6 inches in diameter going through its length. There was a space of about 6 inches between the bottom of the tube and the outer casing. Many persons opposed the new plans. The Boulton and Watt low pressure engine did not work well with the high steam, and the water rose in the mine workings. Captain Trevithick, seeing that he was being swamped, received permission from the mine managers to dismiss the old engine hands and employ his own staff. Captain Jacob Thomas was the man chosen to put things right. He never left the mine until the engine worked better than ever before, and forked the water to the bottom of the mine. Before that time the average duty in the county by the Boulton and Watt engines was seventeen or eighteen millions, and in two or three years, with Trevithick's boilers and improvements in the engines, the duty rose to forty millions...

not everywhere of course, but at those few mines which were studying economy.

His next job, it appears, was the erection at the beginning of 1812 of a small pumping engine at Wheal Prosper in the parish of Gwythian. Fortunately we have a very full account and sketch of this engine from memory by Richard Hosking. It is as follows:[3]

I am happy to be enabled to give you in detail an account of the Wheal Prosper Engine. It was the first Engine I ever worked, therefore the occurrances are fresh in my memory.

[1] The proposals, and the trials on which they were based, are described in letters to Giddy, *Enys Papers*, dated Camborne, 1806, March 4th and March 21st.

[2] *Life*, II, 157.

[3] *Enys Papers*. Letter, R. Hosking, Perran Foundry, to J. S. Enys, 1843, Jan. 17th.

In answer to your questions I beg to say—

1. The Engine was erected early in 1812 (Spring)

2. by Capt. R^d Trevithick, as sole Engineer.

3. The castings were made (I believe at Mr Harvey's, Hayle), wrought iron by Jos. Vivian, Roseworthy.

4. Cylinder 24 ins Diam^r, 6 ft Stroke, equal Beam.

5. Two valves in the top nozzle, 7 inches diameter (common).

6. Cylindrical Boiler, 24 ft long from 5 ft 6 in to 6 ft diameter, with fire Tube in the usual way, except a little flattened on the upper side, so as to give more steam room in the Boiler.

Fires, under and over, three times the length of Boiler.

7. The Engine worked more expansively than any other Engine I have seen, except Taylors at United Mines. Should say cut off at $\frac{1}{5}$, or even less.

8. The Nozzle was small in proportion to the valves and I should think the clearance between the piston & cover did not exceed 3 inches.

9. Main beam under the cylinder, equal length both ends.

10. 20 ftm. 6 inch plunger; 20 ftm. 10 inch D^o; 10 ftm. 12 ins. Bucket. This was her extreme load & I am not prepared to say the extent of her expansion while working thus. The greatest expansion was when her load was from 5 to 6 lbs pr. square inch on the piston.

11. The steam pressure in the Boiler must have been about 40 lbs pr inch. I very well remember assisting Capt Trevithic to put on a weight on the saftey valve, which was a piece of Pump [i.e. pipe] from 3 to 4 cwt; it stood vertical on the valve with a guide at the top the valve being about 3 in. diam^r.

12. There was no throttle valve of any description between the boiler & nozzle; therefore the steam acted on the piston, (being admitted through a 7 inch valve into a 24 inch cylinder) with its full force. The Engine stood indoors between strokes, with the cylinder full of steam, in order to keep it as hot as possible. There was no steam case, but cloathed with straw or Reed and plastered with lime.

The Air Pump was of the plunger description; pole 12 inches diamtr. 3 ft stroke situated within the House; condenser cloase to the bottom of the cylinder.

No doubt the air pump was situated thus for discharging at the downstroke of the Piston.

The Engine was made and erected pr. Contract, viz. Capt Trevithick receiving a certain sum of money for Building the House, providing and erecting a new Engine of the dimensions before given.

As to the duty I can say nothing further than that Capt Trevithick was well pleased with her Performance when she had a light load, but when she became overburthened, she was considered *by some* a bad Engine, but this I should say is no criterion. I never knew an Engine do well when overloaded. I can only say she was put up on speculation, that is an experimental Engine, and the first condensing Engine wherein steam was worked to that degree of pressure.

That the engine was experimental is supported by the fact that the cylinder came from Relistan Mine, as the Account Books show.

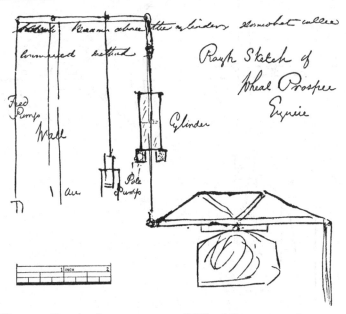

Fig. 19. Sketch of arrangement of Wheal Prosper engine, 1811.
From the *Enys Papers*

Mr Enys called on Mr Hosking the day following this letter and got from him the sketch (reproduced in Fig. 19). Mr Enys, it should be mentioned, was engaged at this time in collecting material for Dr Pole, who was preparing his work on the *Cornish Pumping Engine* and was visiting Cornwall for the purpose.

We have given the above letter *in extenso* because of the importance of this engine, which in the words of Dr Pole[1] "appears to have been the first Cornish engine ever erected; that is the

[1] *Cornish Pumping Engine*, 1844, p. 51.

first condensing engine working with high pressure steam expansively and having the present form of boiler". He said "high pressure steam" advisedly, because Watt had worked expansively with low pressure in 1781. The authors are of the

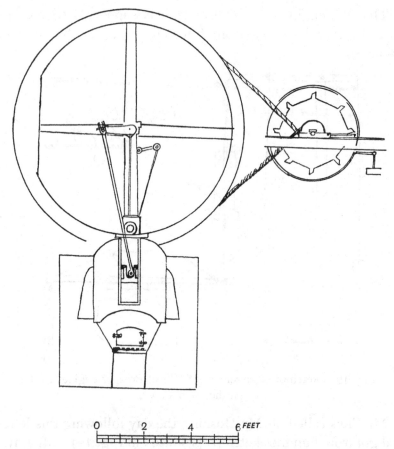

Fig. 20. Trevithick's thrashing engine at Trewithen, 1812.
From the *Enys Papers*

opinion unequivocally that the Wheal Prosper engine was the first Cornish engine. This is no small claim to make on Trevithick's behalf, but the authors do so in justice to a man who never claimed it for himself.

Having got the idea on the brain of expansive working of steam, he proceeded to embody the principle in his portable

PLATE XIV. TREVITHICK'S HIGH-PRESSURE ENGINE AND
BOILER, USED FOR THRASHING, 1812

Courtesy of the Science Museum, South Kensington

rotative engine, suitable for agriculture and other purposes. To do so and yet retain its simplicity, he made the engine single-acting with a 3-way-cock as before, actuated by a cam, by which means he could obtain any cut-off without further complication of the valve gear.

The first of such engines, with its boiler, he built for Sir Christopher Hawkins, Bart., M.P., of Trewithen, early in 1812, for thrashing corn, replacing the cattle mill previously used (see Plate XIV). Trevithick found that 2 bushels of coal costing 2s. 6d. did the work of four horses costing 20s. Three independent observers gave a testimonial[1] to its performance. The engine continued in use till 1879, when it was taken to London; fortunately it is still preserved in the Science Museum, South Kensington. It was so successful that, as a result, a second engine for Lord Dedunstanville was made for work at his seat, Tehidy Park. This also had a long life; only the boiler has survived, to find a resting-place alongside the preceding. Agricultural engines were made for Mr Kendal of Padstow, for Mr Jasper of Bridgnorth and for other persons, as is evident from his correspondence with Rastrick.[2]

In 1812, Trevithick was, according to his son, advertising these engines for thrashing and grinding corn, sawing wood and such like. They weighed 15 cwt. and cost £63. He contemplated, if not actually made, engines that were self-propelling, e.g. in a letter to Sir Christopher Hawkins he mentions:[3]

I am now building a portable steam whim on the same plan [i.e. as that of the thrashing engine] to *go itself* from shaft to shaft; the whole weight will be about 30 cwt and the power equal to twenty-six horses in twenty four hours.

About a year later Trevithick wrote to Rastrick:[4]

I wish you to finish that engine with boiler [i.e. the one to be sent to Exeter], wheels, and everything complete for ploughing and thrashing, as shown in the drawing, unless you can improve on it. There is no

[1] *Enys Papers. Life*, II, 38. Copy of the original.
[2] *Life*, II, 40.
[3] *Life*, II, 172. Letter dated Camborne, 1812, March 10th.
[4] *Life*, II, 55. Letter dated Camborne, 1813, Jan. 26th.

doubt about the wheels turning around as you suppose, for when that engine in Wales travelled on the tramroad, which was very smooth, yet all the power of the engine could not slip around the wheels when the engine was chained to a post for that particular experiment.

A drawing by Trevithick has been preserved—the only clue to the date of which is the water-mark 1813, because it has "neither name, date nor scale"—of a machine which, if not made, shows what was passing through his mind with respect to ploughing by steam (see Fig. 21). The machine is mounted on a timber frame with four wheels, obviously for being hauled to and fro by a steam engine. The back axle by gearing gives motion to a shaft on the axis of the machine. A wheel on this shaft has twelve shovels on its rim. Ahead of the shovel wheel is a beam carrying seven coulters to cut the soil into longitudinal strips, and a curved knife-edged plate below them completes the detachment of the soil, so that the rotary shovel can throw it sideways.

Belonging to this period we have an interesting letter[1] to Robinson and Buchanan, a firm of brewers in Londonderry, who sent an enquiry as to utilizing a stream of water to give them power. Trevithick advised a water-pressure engine. He stated as to pressure of steam: "I do not exceed 20 lbs. finding under this pressure the piston will stand six or eight weeks, and the joints remain perfect and no risk of bursting the boiler, it being made of wrought iron and proved by pressure before sent off". He discusses also the use of exhaust steam for process operations, i.e. to heat the mash tun. What resulted can best be stated in Trevithick's own words:[2]

I have yours of the 20th November. The letter you directed for Truro never came to hand. I find by your letter that you have been trying to put into practice the hints I gave you about the chain and buckets, and that you expect it will answer if properly executed. You are not the first that has picked up my hints, and stuck fast in their execution. I make it a rule never to send a drawing until I have received my fee, and when you remit to me fifteen guineas I will furnish

[1] *Life*, ii, 12. Draft dated Camborne, 1812, March 5th.
[2] *Life*, ii, 16. Draft dated Camborne, 1812, Dec. 5th.

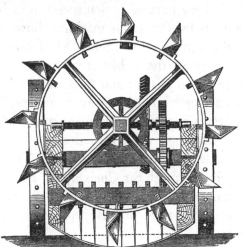

Fig. 21. Trevithick's steam cultivator, 1813. Drawing
prepared by his son, from his father's original

you with proper drawings and directions to enable you to make and erect the machine.

Sir Christopher Hawkins was so pleased with his thrashing machine that he communicated his opinion to Sir John Sinclair, Bart., originator and first President (1793–1806) of the Board of Agriculture, then a very live Department. Sir John wrote to Trevithick asking for drawings and particulars. This the latter supplied in a letter dated from Camborne, March 26th, 1812, through the intermediary of Sir Christopher Hawkins, offering at the same time to send an engine to the Board as a sample. The letter is concerned also with proposals for steam navigation, which will be considered later.

Sir John replied on April 4th, informing Trevithick that he had sent the suggestion for propelling vessels by steam to the Navy Board and asking for further particulars of the thrashing engine. Trevithick in his reply then proceeded to unburden himself of a "few wild ideas" of his. They were not wild at all, being clear statements based on what he had done with loco-motive and stationary engines, as to what could be done by steam as compared with manual labour, and registering his conviction "that every part of agriculture might be performed by steam...and all by the same machine, however large the estate". Trevithick makes this offer: "My labour in invention I would readily give to the public if by a subscription such a machine could be accomplished and made useful".[1]

A Government Department, however enterprising, was not the body to initiate expensive experiments, and we hear of nothing further resulting in this direction.

But another outlet for engines occurred to Trevithick as the result of an enquiry through Sir Christopher Hawkins, and that was to work sugar mills in the West Indian plantations. It will be recalled that such an engine is represented in the patent drawing of 1802.

During the years 1812–1816 he had many enquiries from the West Indies, and it is more than probable that he sent several

---

[1] *Life*, II, 43. Letter dated Camborne, 1812, April 26th.

engines there. One was actually constructed for a Mr R. W. Pickwood of the Island of St Kitts, West Indies. Trevithick gave the order for its construction to Hayle Foundry. As, however, there arose a dispute among the partners, Blewett, Harvey and Vivian (Trevithick's old partner who had now joined the concern), the foundry was stopped by an injunction of the Lord Chancellor, and the engine was eventually made by Hazledine, Rastrick and Co. at Bridgnorth. The delay, however, caused it to be too late for shipment and it ultimately found its destination in the Mint at Lima. However, in a letter written by Trevithick on March 18th, 1814, he says that he "will get an engine ready for Mr Pickwood by the time mentioned of the power and for the price fix'd in our former corrispondence".[1] This was after the engine originally intended for Pickwood had been sold for the purpose of sending it to Lima.

During this dispute at Hayle, Trevithick was at Henry Harvey's house, when[2]

Blewett sent a handsome silver tea-pot to Miss Betsy Harvey, who kept her brother's house.... Trevithick was sitting with them when the box was brought in and opened. Mr Henry Harvey was indignant at Mr Blewett sending a bribe or make-peace to his sister, and threw the silver tea-pot under the fire-place. Trevithick however quietly picked it up, pointed out the dinge it had received, wrapped his pocket-handkerchief around it and saying if it causes bad feeling here, it will do for Jane, marched away home with the pot.

It is, perhaps, not too much to claim for Trevithick the credit for the introduction of the high-pressure engine into the West Indies, through Rastrick, first at the Bridgnorth and afterwards at the Stourbridge Works. Joshua Field, in his *Diary* in 1821, says of the latter works:[3] "They make all their own steam engines and carry on an extensive engine trade for the West Indies".

In November of 1812, Trevithick was building an engine to the order of Mr James Green, for the purpose of draining the foundations of Exeter Bridge, then under construction.[4] The

---

[1] *Draft Letter Book.*   [2] *Life,* ii, 52.
[3] *Trans. Newcomen Soc.* vi, 28.   [4] *Life,* ii, 22, 29.

engine was promised for February, 1813. It was constructed at Hayle Foundry and was for the low lift of 12 ft., the pump bucket being 10 in. diam. While carrying out this job, he learnt from Mr Green that a large amount of stone was to be quarried for the Plymouth Breakwater.

It should be mentioned that various schemes had been mooted for the protection from southerly gales of the anchorage in Plymouth Sound by breakwaters or piers, and this culminated in John Rennie being asked by the Admiralty to report on the various proposals, which he did in April, 1806. He recommended a breakwater and submitted plans. Five years of controversy over the proposals and plans ensued, but eventually, in June, 1811, an Order in Council was made to carry out the work. Suitable stone was obtainable at Oreston up the Catwater about three miles distant, and 25 acres were bought, quarries were opened out, railways laid down and wharves constructed.

Trevithick, seeing an opportunity of using his engine in connection with the construction of the breakwater, wrote to his brother-in-law to get to know something about the possibility of doing so, and received the following reply:[1]

I am in receipt of yours of the 22nd inst. Mr Giddy informs me that Mr Fox and Mr Williams are to have 2s 6d per ton for making the breakwater at Plymouth, and he considers that they can do it for 2s, which he thought would give them 50,000l. profit. If you meet those gentlemen, I have to caution you not to LEARN THEM anything until you make a bargain, as I know Mr Williams will endeavour to learn all he can and then you may go whistle.

If 6d. per ton will give 50,000l. profit, a halfpenny per ton would give upwards of 4000l. Would they agree to give you that for your labour only? However, this will depend in a great measure on the time it will take in doing. If it takes eight years it would be 500l. a year for you (according to Mr Giddy's calculation).

Before he received his brother-in-law's letter, however, Trevithick had been to Plymouth, in company with his friend Rastrick, to see for himself. We learn about this trip in a

[1] *Life*, ii, 23. Letter from Henry Harvey, dated 106, Holborn Hill, 1812, Nov. 26th.

gossiping letter from the latter to Goodrich.[1] Rastrick had travelled from Stourbridge on horseback, crossing the Bristol Channel into the West country. He says: "I... went to Redruth and Cambourn... Trevithick lives within 2 miles of Cambourn", but he was not at home that day.

On my return next Day I found Trevithick at home. I spent several days with him and he went with me to see several people who gave me orders—I mentioned to him that I was going to see·the Breakwater at Plymouth. He thereupon agreed to accompany me.... Trevithick and I set out and rode along the Southern Coast of Cornwall to Plymouth. I delivered your Letter to Mr Mitchall who conducted us thro' the Dock Yard, but I was by no means so well pleased with it as I was with Portsmouth. We now went to Oreston Quay to see the Manner in which the Business was conducted there of getting and Shipping the Stone—we found the Principal Quay was rendered quite useless as it had sunk according to the information we had *recieved* about 20 feet not perpendicular but by inclining in such a Way that should they continue to build on it, the Foundation will probably be uppermost as it seems to turn thus [sketch omitted]. They were building on it when we were there altho' it is evident that no *vessell* can come along side of it, since the Mud or Soil is hove up in such a way as not to allow a sufficient depth of Water. The Method in which they get the Stones and convey them to the Quays; the Construction of the *Vessells*, the way of Loading and discharging is in my Oppinion, the *verry verry worst* that could be thought of by the greatest Bungler that ever was. You are I suppose acquainted with the Construction of the Vessells which carry the Stones; they have a lower and upper Deck with two Rail Roads on each Deck paralel with each other and I believe there are seven or Eight Carriages on each Rail Road these Vessells are about 120 Tons Burden but they do not carry above 60 Tons of Stone for if I recollect rightly I calculated that the unnecessary Lumber, Rail Roads, Carriages &c. &c. on board was about half the Burden of the Vessell and really the Length of Time in unloading the Vessell beyond all bounds—There are a variety of Ways which occurred to Trevithick and me and which might enable the Contractors to get Money by their Contract provided they were allowed to adopt the Method that could be proposed, as it is I am told the Contractors are loosing a great Deal and must continue so to do—On the whole it is the worst conducted Public Work I ever saw.

[1] Goodrich Papers, Science Museum, South Kensington, quoted in *Trans. Newcomen Soc.* IV, letter dated 1812, Nov. 15th.

Obviously Trevithick shared his companion's opinion as to the scope there was for improvement. He at once started experiments on boring limestone mechanically and communicated his results and opinions to Mr Robert Fox of the firm of Fox, Williams and Co., the contractors for the breakwater, neglecting his brother-in-law's worldly-wise advice not to "learn them anything until you make a bargain".

The letter so teems with practical suggestions that we give it in full:[1]

Since I was at Roskrow I have been making trial on boring lumps of Plymouth limestone at Hayle Foundry, and find that I can bore holes five times as fast with a borer turned round than by a blow or jumping-

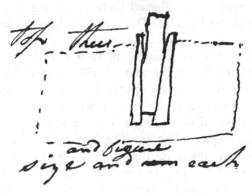

Fig. 22. Plug and Feathers for breaking up stone, 1813

down in the usual way, and the edge of the boring bit was scarcely worn or injured by grinding against the stone, as might have been expected. I think the engine that is preparing for this purpose will bore ten holes of 2½ inches in diameter 4 feet deep per hour. Now suppose the engine to stand on the top of the cliff, or on any level surface, and a row of holes bored, 4 feet in from the edge of the cliff, 4 feet deep, and about 12 inches from hole to hole for the width of the piece to be brought down at one time, and wedges driven into the holes to split the rock in the same way as they cleave moorstone, only instead of holes 4 inches deep, which will cleave a moorstone rock 10 feet deep when the holes are 14 or 15 inches apart, the holes in limestone must go as deep as you intend to cleave out each stope, otherwise the rock will cleave in an oblique direction, because detached moorstone rocks have nothing to hold them at the bottom, and split down the whole depth of the rock. In carrying down a large piece of solid ground the bottom will always be fast, therefore unless it is wedged hard at the bottom of the hole the stope cannot be carried down square. In a hole $2\frac{1}{2}$ inches diameter and 4 feet deep put in two pieces of iron, one on each side of the hole, having a rounded back, then put a wedge between the two pieces, which might be made thus [cf. Fig. 22], if required to wedge tighter at the bottom of the hole than at the top.

If this plan answers, the whole of the stones would be fit for service, even for building, and would all be nearly of the same size and figure. Each piece would be easily removed from the spot by an engine on a carriage working a crane, which would place them into the ship's hold at once. It would all stand on a plain surface, and might be had in one, two, three, or four tons in a stone, as might best suit the purpose, which would make the work from beginning to end one uniform piece. Steam machinery would accomplish more than nine-tenths of all the work, besides saving the expense of all the powder. I find that limestone will split much easier than moorstone, and I think that a very great saving in expense and time may be made if the plan is adopted.

Please to think of these hints and write me when and where I may see you to consult on the best method of making the tools for this purpose before I set the workmen to make them.

Trevithick followed this up with further explanations:[1]

Since I was with you at Falmouth I have made a trial of boring limestone, and find that the men will bore a hole $1\frac{1}{2}$ inch in diameter 1 inch deep in every minute, with a weight of 500 lbs. on the bit. I had no lump more than 12 inches deep; but to that depth I found that having a flat stem to the bit of the same width as the diameter of the hole, twisted like a screw, completely discharged the powdered limestone from the bottom of the hole without the least inconvenience.

From the time the two men were employed boring a hole 12 inches deep, I am convinced to a certainty that the engine at Hayle will bore as many holes in one day as will be sufficient to split above 100 tons of limestone, and would draw that 100 tons of stone from the spot and put them into the ship's hold in one other day. The engine would burn in two days 15 bushels of coal, four men would be sufficient to attend on the engine, cleave the stone, and put it into the ship's hold. I think it would not amount to above 9d. per ton, every expense in-

---

[1] *Life*, II, 27. Draft letter to Robert Fox, dated Camborne, 1813, Feb. 4th.

cluded, but say 1s., which I am certain it will not amount to. Perhaps it may not be amiss to withhold the method of executing this work until the partners have more fully arranged with me the agreement as to what I was to receive for carrying the plan into execution. I do not wish that anyone but your father should be made acquainted with the plan, and have no doubt he will have sufficient confidence in the scheme to adopt it. I shall be glad to hear from you soon, as I intend to go to Padstow in a few days and shall not return under a fortnight.

In writing to Rastrick, Trevithick says:[1]

Messrs. Fox will want a great many engines of that size for the Plymouth Breakwater. They are to provide machinery, with every other expense, and I am to have a certain proportion of what I can save over what it now costs them to do it by manual labour. I think I have made a very good bargain, for if the plan succeeds, I shall get a great deal of money, and if it fails I shall lose nothing. They have engaged with the Government to deliver 3,000,000 tons, for which they have a very good price, even if it was to be done by men's labour. I hope I shall get the engine soon on the spot, and will then let you know the result.

On March 14th he wrote to Fox that "the engine with the boring apparatus for Plymouth remains at Redruth", and on May 20th to Rastrick: "The engine for Plymouth will be put to break the ground as soon as I can find time to go up there". We do not know the final result, but the Messrs Fox were honourable men and we feel sure that Trevithick was paid for his work. His plans, or something like them, must have been adopted, for the price for getting the stone and dumping it was shortly afterwards reduced from 2s. 9d. to 1s. per ton.

Trevithick did not patent the single-acting expansive engine himself but he discussed with Rastrick the propriety of doing so. In a long letter to the latter, March 12th, 1813, Trevithick says:[2] "There is scarcely a post but I am plagued with orders for engines but I am tired answering them as I cannot meet their orders for want of castings....I have no doubt a patent for this plan would stand".

Probably what decided him was the fact that he had not obtained

---

[1] *Life*, ii, 55. Draft letter, dated Camborne, 1813, Jan. 26th.
[2] *Draft Letter Book.*

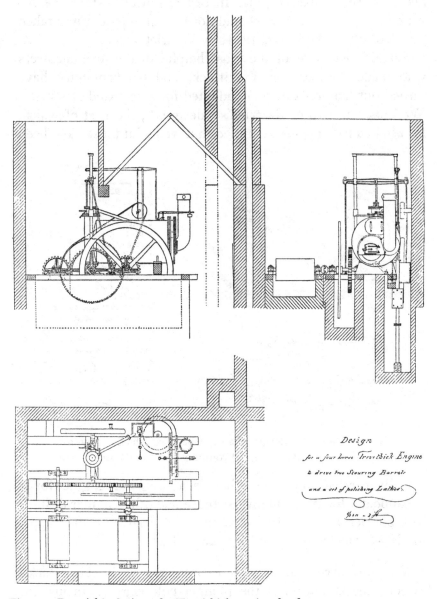

Design
for a four horse Trevithick Engine
to drive two Scouring Barrels
and a set of polishing Lathies

Fig. 23. Rastrick's design of a Trevithick engine for factory purposes, c. 1813.
From the *Goodrich Papers*

his discharge in bankruptcy. In order, however, that the advantages of the invention should not be lost, a patent was taken out, No. 3799, April 1st, 1814, by Rastrick alone.[1]

Rastrick was one of the most helpful of the manufacturers who made engines for Trevithick, and his firm must have turned out hundreds of them adapted for every kind of service. We illustrate two of Rastrick's "designs", the first of which, for use in a factory, is so detailed that we feel it must have been

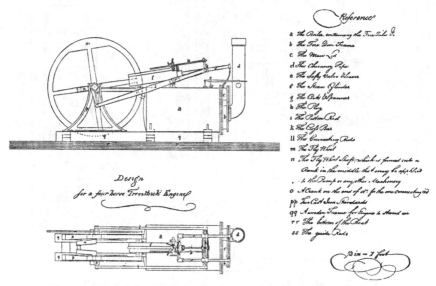

Fig. 24. Rastrick's design of a diagonal Trevithick engine for a steamboat, c. 1813. From the *Goodrich Papers*

actually carried out; the other is a diagonal engine for a steamboat. Rastrick submitted these and other similar designs to Goodrich about 1813 as examples of what he could supply for H.M. Dockyards (see Figs. 23 and 24).

We now arrive at another definite period in Trevithick's career—the invention of the plunger-pole engine. In just the same way as he had reversed the plunger pump by using water under pressure to actuate the plunger and so produced the water-pressure engine, so now he substituted steam for water to act on

---

[1] Fully discussed in *Trans. Newcomen Soc.* VII, 42.

the plunger and converted it into the single-acting plunger-pole engine. The pole case, acting as the cylinder, stood on beams across the shaft, thus somewhat recalling the Bull engine; a cross-head on the pole sliding in guides was connected by side rods to a cross-piece below with the pump rod hanging from it. There was a condenser and air pump worked from a balance-bob.

The first plunger-pole engine was erected at Wheal Prosper in 1812; in fact it must have been in hand at the time he erected the first Cornish engine at the same mine, as described above. The pole was 16 in. diam. and the stroke 8 ft. There were two boilers 3 ft. diam. by 40 ft. long externally fired, the steam pressure being 100 lb. to the sq. in. Captain Samuel Grose, then a pupil of Trevithick's, was in charge of the erection.[1]

It is necessary, in order to understand what ensued, to know that Arthur Woolf, who had been employed at Meux's Brewery,[2] had come back from London to Cornwall, in order to exploit in the mines there the compound engine and sectional boiler that he had recently patented. Woolf was consequently in rivalry with Trevithick. Either of their engines was superior to the Boulton and Watt, and adventurers who wanted new or improved engines were not a little nonplussed to choose between them. One of a number of stories of this rivalry is told by Samuel Grose: James Oats was at work on the boilers for Wheal Prosper when "Woolf came into the yard and examined them.[3] 'What d'ost thee want here?' asked Oats. 'D—n thee I'll soon make boilers that shall turn thee out of a job!' was Woolf's reply", meaning by that that his patent sectional cast-iron boiler would do away with the necessity for the skill of the boilersmith. Woolf was a rough customer!

It is the engine at Wheal Prosper that is referred to by Rastrick in his letter of November 15th, 1812, when he writes:

I went to see an Engine put up by Trevithick, which works with less coal than any Engine I know of. It is a single Power Engine with

---

[1] *Life*, II, 70, where an illustration is given.

[2] In the *Enys Papers* we have correspondence of Woolf and Trevithick with Giddy, showing that Trevithick tested Woolf's engine there on March 8th, 1808.   [3] *Life*, II, 111.

an Air Pump and Condenser, only two Nozzle Valves, but the principal Advantage it has is by working with a Steam of high elastic Power and shutting off the Steam at one third the Stroke, it expands for the other two thirds of the Stroke, is then so far lowered in Power as not to exceed the Power of the Atmosphere and of Course becomes verry easily condensed.

That visit is recalled by Trevithick in his letter to Rastrick the following January:[1] "That new engine you saw near the sea side with me is now lifting forty millions 1 foot high with 1 bushel of coal, which is very nearly double the duty that is done by any other engine in the County".

Trevithick now decided to patent the engine and did so on June 6th, 1815 (Specification No. 3922). The description in the specification is, as usual with Trevithick, remarkably clear, but it is unnecessary to reproduce it, as the engine has been already described. Nor is it necessary to detail the other engines on this plan erected by Trevithick except the last—that at Herland Mine, 1815, the one of which he was most proud. We can best give the description of it in his own words:[2]

Woolf's engines is Stop'd at Herland and I have orders to proceed. A great part of the work is finished for them and will be at work within two months from the time I began. I onely engage that this engine shall be equal to a B & Watts 72 In. single, but it will be equal to a Double 72 In Cylinder. It is a Cast-iron plunger pole over the shaft of 33 Inches diamr, 10-feet Stroke. The boiler is two tubes, 45 ft Long, each 3 ft Diamr, ½ Inch thick of wrought iron, side by side, nearley horizontal, onely 15 Inches higher at the steam end of the tubes, to let the steam free to the steam-pipe. There is two four Inch valves, one the steam-valve, the other the disscharg valve. I have made the plunger case and steam vessell of wrought Iron ¾ Inch thick. The steam vessel is 48 In Diamr. The plunger stands on beams over the shaft, with the top of it to the level of the surface, with a short T-pice above the plunger pole, and a side rod on each side, that comes up between the two plunger beams in the shaft; which does away [with] the use of an Engine beam, and the plungers does away with the

[1] *Life*, II, 55. Letter dated 1813, Jan. 26th.
[2] *Enys Papers. Life*, II, 80. Letter to Giddy, dated Camborne, 1815, July 8th.

use of a ballance beam. The fire is under the two tubes, and goes
back 45 feet under them, and then returns again over them, and then
up the chimney....

This engine, every thing new, house included, and sett at work at
the surface will not exceed £700 and 2 month [will be sufficient for]
erecting. The Engine of Woolf's, at Whl Vor, which is but two thirds
the power of a 72 In Cylinder, single power, cost £8000, and was two
years errecting.

Trevithick met with endless difficulties over this engine, not
by any means all connected with its technical side. He had sent
so many orders to Bridgnorth that he had created unpleasant-
ness among the founders who had previously worked for him—
Joseph Price of Neath Abbey, Henry Harvey of Hayle, Andrew
Vivian, and the men at Perran Foundry. They exerted influence
either as adventurers themselves or with the adventurers. When
Trevithick had at last got the castings, it was found that the pole
was not truly turned and the work had to be done over again.

Trevithick wrote a long letter of explanation to Mr Wm.
Phillips, "the principal of the London adventurers" in Herland:[1]
"This is very uncivil treatment in return for inventing and
bringing to the public at my own risk and expense, what I believe the country could not exist without". He complained also
to Rastrick about the workmanship: "the whole job is most
shamefully fitted up";[2] space will not permit of quoting more of
either. There was difficulty, too, in getting the engine to start.
Henry Phillips in his reminiscences,[3] 1869, gives a graphic
account:

I was a boy working in the mine, and several of us peeped in at the
door to see what was doing. Captain Dick was in a great way, the
engine would not start; after a bit Captain Dick threw himself down
upon the floor of the engine-house, and there he lay upon his back;
then up he jumped, and snatched a sledge-hammer out of the hands of a
man who was driving in a wedge, and lashed it home in a minute.
There never was a man could use a sledge like Captain Dick; he was

[1] *Life*, ii, 88. Letter dated Penzance, 1815, Dec. 13th.
[2] *Life*, ii, 89. Letter to Hazledine, Rastrick & Co., dated Penzance, 1815,
Dec. 23rd.
[3] *Life*, ii, 90.

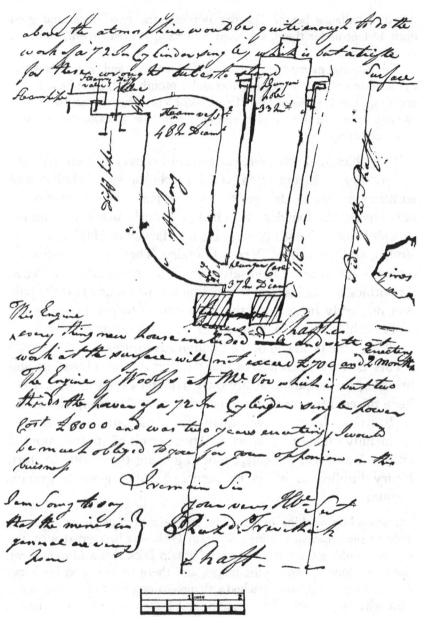

Fig. 25. Trevithick's plunger-pole engine, 1815

as strong as a bull. Then, he picked up a spanner and unscrewed something, and off she went. Captain Vivian was near me, looking in at the doorway; Captain Dick saw him, and shaking his fist, said: "If you come in here I'll throw you down the shaft". I suppose Captain Vivian had something to do with making the boilers, and Captain Dick was angry because they leaked clouds of steam. You could hardly see, or hear anybody speak in the engine-house, it was so full of steam and noise.

On February 11th, Trevithick wrote to Giddy a most glowing account of the performance[1] of the engine, giving particulars; it appears that it commenced work at the beginning of February, that it was actually 33 in. diam. by $10\frac{1}{2}$ ft. stroke, 12 strokes per min., cutting off at $\frac{1}{4}$, and expanding to below atmospheric pressure. He had tested the boiler to 120 lb. and intended to work at that pressure:

I have offer'd to deposit a thousand pounds to £500 as a bet against Woolf's best engine and give him 20 Mill[ns], but that party refuses to accept the chalange. I have no doubt but by the time she is in fork that she will do 100 mill[s].

He asks Giddy to calculate the duty, which he does, and finds it to be $57\frac{3}{4}$ million.[2] In this letter there is this striking observation:

Some recent Experiments made in France prove, as I am told for I have not seen them, that very little Heat is consumed in raising the Temperature of Steam, and if this is true, of course there must be a great saving of Heat by using steam of several atmospheres strong, and working expansively through a large portion of the cylinder.

Trevithick also wrote to Phillips:[3]

The engine continues to work well. Every person that has seen it, except Joseph Price, A. Vivian, Woolf, and a few other such-like beasts, agrees that it is by far the best engine ever erected. Its performance tells its effects, in spite of all false reports.... I could not help.... threatening to horsewhip J. Price for the falsehoods that he with the others had reported.

[1] *Enys Papers. Life*, II, 91. Letter, Trevithick to Giddy, 1816, Feb. 11th.
[2] *Enys Papers. Life*, II, 94. Letter dated East Bourne, 1816, Feb. 15th.
[3] *Life*, II, 96. Letter dated Penzance, 1816, March 8th.

Due to this underhand work, independent arbitrators were appointed,[1] who reported that it did a duty of 48 millions and they had no doubt but that the engine will preform above 60 millns before she get to bottom of the mine.... The engine work^d 9¼ [strokes] pr. Mt. with 2 Bus^{ls} of Coals pr Hour for the whole time, 10 ft Stroke, two lifts of 4½ In. buckett, making 43 fm. & 26 fm. of 6 In. house water. The steam was from 100 to 120 p^d to the inch, the valve open while the plunger pole ascended 20 Inches and then went the remainder of the 10 ft Stroke expancive. It went exceeding smooth & regulier, the most so that I ever saw.

Not content with having brought out the pole condensing and the pole non-condensing engines, Trevithick had a further idea, that of adding a pole case to existing Watt engines. This necessitated, of course, supplying new high-pressure boilers and in fact meant compounding the engines. In this Trevithick became associated with William Sims, a leading engineer of the Eastern Mines, who, acting for the Williams family, one of the largest adventurers in that district, bought a half share in the pole patent as regards Devon and Cornwall. The agreement was signed in fact only on the eve of Trevithick's departure for Peru. The two engineers thus compounded engines at Wheal Chance (45 in.) and Treskerby (58 in.) with considerable increase in duty. Sims alone later compounded three more, i.e. at United Mine, Williams (65 in.), Wheal Damsel (42 in.) and United Mines, Poldory (63 in.).

On June 28th, 1816, Trevithick wrote:[2] "I have now eleven engines on this plan at work or nearly finished", presumably by this meaning all varieties of the plunger pole.

The engine had serious defects: the large surface of the pole exposed to the atmosphere caused cooling, which was greater the greater the pressure; the middle of the pole wore more than the ends, causing leakage, and the packing was difficult to keep tight round the pole. Probably Trevithick would have been the first to have dropped the pole engine if he had stayed longer in

[1] *Enys Papers. Life,* II, 99. Letter, Trevithick to Giddy, Penzance, 1816, April 2nd.
[2] *Draft Letter Book.*

Cornwall. A significant fact is that Woolf dropped his compound engine and his sectional boiler. Henceforward development went on the lines of the Cornish engine—and Trevithick's boiler—with higher pressures and earlier cut-off, as we shall see in a later chapter.

We have referred[1] to Trevithick's experiments in propelling a canal barge by paddle-wheels, and to his proposal in the patent of 1808 to propel a vessel by a central paddle-wheel in an airtight casing. He reverted to the subject of steam navigation in 1812. In a letter to Sir John Sinclair[2] he says: "This is a subject to which I have given a great deal of thought, as being a thing of immense magnitude and value, if it can be made to be of general utility". His idea seems to have been to use a screw propeller, but we have no details. However, he took up the matter again in 1815. Just as he did so, he was struck by another idea—that of an engine on the principle of Heron's aeolipile, producing rotary motion by the reaction on the atmosphere of steam issuing from tangential orifices. His main idea was to get a light motor. Like so many prolific inventors, Trevithick made the mistake of combining several new ideas in one experiment, and he now tried these two ideas together and characteristically on a considerable scale. This was done at Bridgnorth, as appears from a report of what he was doing:[3]

Yesterday I fix'd the pumps to the new Engine which goes exceeding well. The bucket is 24½ Inch diamr, 18½ feet high, Stroke 3 feet, work'd 15 St. pr. Mt, with aboute 28 pounds of coals pr Hour. The arms is 15 feet from point to point, but onley one of them give steam. The[y] run 20 rounds for one Stroke of the engine which is worked by a strap, like a comn lathe. The steam was about 100 lb. to the inch. The opening in the arm was about ½ Inch long by ¼ Inch wide. I intend to work the steam much higher. The fire place is one tube in another. The fire-tube is 17½ Inches diamr; the outer tube 24 Inches stands perpender 9 feet high.

[1] Chap. III, p. 79.

[2] *Life*, I, 343, 344. Letter dated Camborne, 1812, March 26th.

[3] *Enys Papers. Life*, I, 364. Letter to Giddy, dated Bridgnorth, 1815, May 7th.

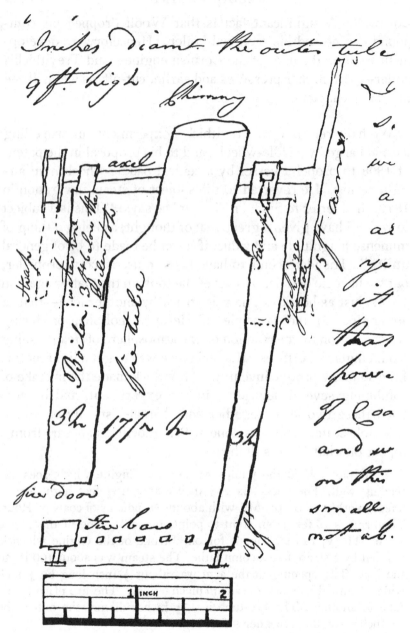

Fig. 26. Trevithick's recoil engine and boiler, 1815

I shall continue to make several experiments, and will give you the result, but as yet cannot say much aboute it, as it was late yesterday before I finish'd it, but I am convinced that it is by far the greatest power for the Consumption of Coals that was ever made, and when I consider the friction on this small machine and the small fire place, where there is not above three gallns of coal, with the work that it do perform, I can scarceley beleive my own eyes. I have wrote by this post to Cornwall to desire the Foxes to send up their Engineer to inspect it. The large engine is begun, according to the drawings. It will be about 35 times the fire side and fire place of this small engine, but I think it will do 50 times the work, becase the friction will be much less on a large than a small machine. The expence will be about £800, and the power will be equal to a 70 Inch Cylinder double power. I shall be able to carry every part of this powerfull engine on my back (boiler excepted). I would be much obliged to you to inform me what speed you think would be the best for the ends of the arms to travel.

The aspect of this engine with arms projecting $7\frac{1}{2}$ feet from the centre of the shaft and whirling round at 300 revs. per minute must have been startling and even alarming; no wonder the workmen nicknamed it "the windmill" and believed it to be intended to throw balls at an enemy.

In the *Enys Papers* there is a memorandum containing particulars of a number of experiments, obviously those made on this first engine. The second engine, mentioned in the latter part of the letter, is further discussed at great length in a letter to Giddy;[1] and Trevithick built a boiler of entirely new construction, which he describes and sketches for Giddy in a letter dated May 16th:[2]

The boilers thats making for this towing engine is all tubes of 3 ft. Diam^r $\frac{3}{8}$ thick with Circular Ends of wrought Iron. There is three horizontal tubes and to each of these horizontal tubes there is three other tubes suspended by them perpendiculear which will all be fix'd or hung up in a large Chimney thus:

Tho' I have call'd it stone work it will be for portable purposes, instead of stonework, an outside Case of Iron luted with fire clay, but

[1] *Enys Papers. Life*, I, 367. Letter dated Hazledine Foundry, Bridgnorth, 1815, May 12th.

[2] *Enys Papers. Life*, I, 370. Letter, Trevithick to Giddy, 1815, May 16th.

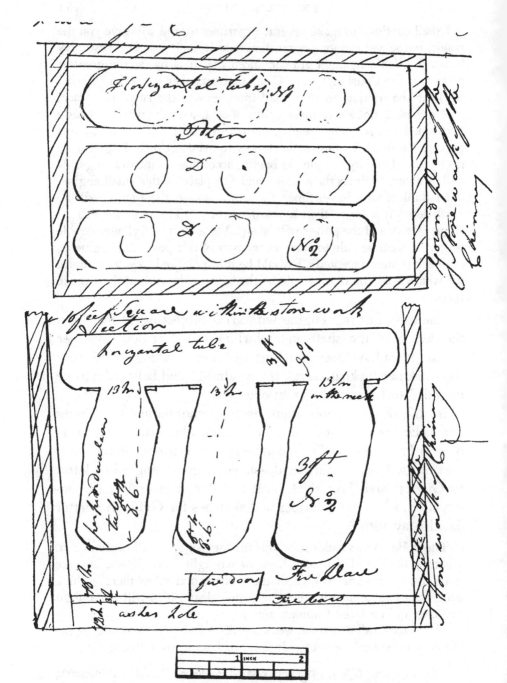

Fig. 27. Trevithick's water-tube boiler, 1815

for fixtures stone work. I am satisfyde that for rotary engines that it will do well, but for engines lifting water the speed must be so much reduced that it will be more expense for that purpose than for machinery, and I have been thinking of a method to use high steam in another way for that purpose, which I send you a sketch of for your inspection. Suppose I work with 200 p^d to the Inch pressure, which the theory of the tubes on the other side will stand according to the practice of proveing the wrought Iron Cables which stands 50,000 p^d to the square Inch of Iron, above 2,000p^{ds} to the Inch on the tubes. Therefore the pressure that I intend to work with will be onely 10 pr Ct of the real strength.

Suppose a common plunger pice and pole on the usual way to be put over the shaft, or a little below the surface, to stand on a beam of wood across the pit in the usual way, and on each side the Iron plunger pole side rods to connect to the shaft rods below and to be fill'd with could water, and from the bottom of the plunger pice a small pipe to convey the water forward and backward to a steam vessell which would pass and repass every stroke by the steam forceing on the surface of the water in the steam vessell. In this way there would not need any engine house or beam, and the plunger and pole and fixing with pipes and everything belonging (boiler excepted) for an Engine of the power of 63 Double [would] not [cost] above £250 or £300, becase a 20 In plunger pole would be sufficient for that power. By this plan the expancion of steam might be made use of. Would the steam on the surface of the water heat it so as to heat it above the boiling point and then imbibe heat and again disschardge it, as the pressure of the steam is thrown on and off? Or would it not answer to put a plate of cork on the face of the water? By this plan a piston could be keep'd tight to any pressure, and the heavy rods would take the advantage of the expanction of Strong steam, and I cannot sea but that the engine would be compleatly manageable by two three Inch valves which would be Verry easely work'd and the whole expence of the machine would not excead £1,000 for the power of a 63 Double power.

What Trevithick accomplished with his rotary engine is described by Richard Preen in reminiscences taken down in 1869:[1]

Mr Trevithick made another kind of engine called the Model, some people called it the Windmill, and said it was intended to throw balls against the French. There were two great arms, each of them 10 or

[1] *Life*, I, 367.

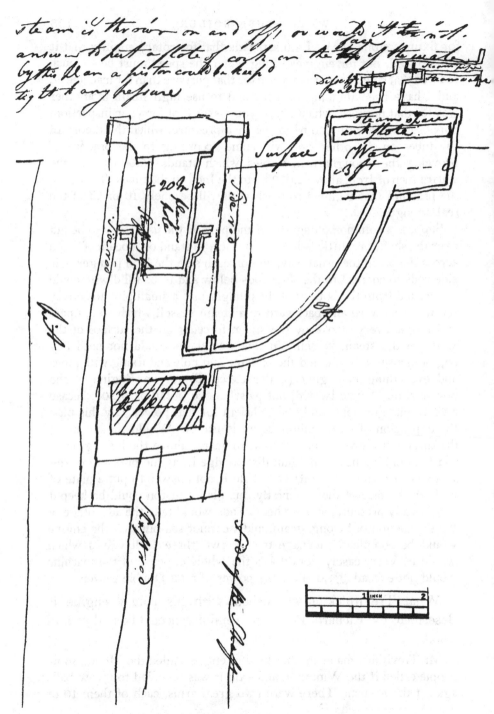

Fig. 28. Trevithick's plunger-pole engine with water diaphragm, 1815

12 feet long, placed opposite one another on a hollow shaft or axle, which had a nozzle in it. When steam was turned on it puffed out at the ends of the arms and they went around like lightning, with a noise like shush! shush! so then it was called by that name. This engine was made just before Mr Trevithick went to South America. He did not know what to do with it, so gave it a present to Jones, the foreman in the works.

This whirling engine or recoil engine, as it was sometimes called, was included in the pole patent of June 6th, 1815 (Specification No. 3922) in these words:

I claim as my invention the various forms of causing steam of a high pressure to pass from a boiler through one or more perforated revolving arms, and to spout out against the atmosphere from an aperture at the side near the extremity of the arm or arms.

As stated, the claim was not novel, since he describes no improvement on the aeolipile. Possibly he became convinced of this, because he wrote, long afterwards:[1]

When I made the werling engine several years ago, I found that the great loss of power was ocasioned by not being able to drive the spouting arm at the extream end not above 200 feet per second, which bore a small propotain to the volosety of the spouting steam, and it's allmost impossiable to obtain sufficient speed to gain the whole of the advantages unless there is some plan that will allow the steam to continue its speed and at the same time to reduce the speed of the machine and give the advance of power on the machine in return for the loss of speed.

The screw was also included in the same patent of 1815. He describes it as "a worm or screw, of a number of leaves placed obliquely round an axis similar to the vanes of a smoke jack". It was "to revolve with great speed" and was to be "in a line with the required motion of the ship or parallel to it".

In July Giddy wrote:[2]

I hope that your experiment with Recoil Engine for moving Boats is in progress. Nothing occurs to me respecting them, except that

---

[1] *Enys Papers.* Letter to Giddy, dated Highgate, Lauderdale House, 1832, Jan. 9th.

[2] *Enys Papers.* Press copy, dated East Bourne, 1815, July 18th.

their is not any theoretical limit to the most advantageous velocity for propelling screws.

He goes on to compare screws of different diameters and velocities.

Apparently Trevithick made up his mind as to what to try, for in November he ordered an engine and screw propeller from Rastrick:[1]

Enclosed you have a drawing for the towing engine for London, which you will execute as soon as possible....On receipt of this drawing I wish you to inform me when you can execute it, and also whether you can understand the drawing. I expect that the screw will be too large for the barges to take to London. If so, leave it in two parts, and we will rivet them together in London.

In February of the following year from his account book we learn that he shipped from Penryn to London in the "Pleasant Hill", Stephen Catt, Master, consigned to Mr John Mills, 4, Canton Place, Limehouse, two boilers, 3 ft. diam. by 36 ft. long, together with a "case" made of old boiler plate. Both boilers and the case were made by James Oats. No doubt these were the boilers and the case to be put into the barge. The iron casing was intended, no doubt, to take the place of the brickwork setting used on land, because brickwork was inapplicable on board ship.

There are letters of this date to James Smith, of Limekiln Lane, Greenwich, who was a millwright working under Trevithick's direction in London. Both he and Mills had by this time become shareholders in the Peruvian enterprise to be described later. Smith,[2] in a business letter, says: "Penn called on me a day or two ago and said he was going to Assist Mr Mills in fitting up the towing barge". In another letter dated July 26th, he says:[3]

He [i.e. Mr Mills] has seen the small engine which I have nearly finished, and wishes to try it on the water with your screw. I shall,

[1] *Life*, I, 352. Letter to Hazledine, Rastrick & Co., dated Penzance, 1815, Nov. 30th.
[2] *Life*, I, 354. Letter, J. Smith to R. T., Greenwich, 1816, Feb. 29th.
[3] *Life*, I, 356, 357.

therefore, be obliged to you to say by return of post what you think should be the diameter of the screw for this size engine, and in what way it would be best to work it: perhaps with a universal joint, if the engine will go fast enough this way; but let me have your opinion on this particular. I expect to have it finished in a week, when Mr Mills will be ready with a proper boat for the purpose. He seems to have some doubts of the screw, but says that if this should not answer, the old way with the wheels will do very well.

In a further letter of August 1st, Smith writes: "I...shall proceed with the screw according to your direction but am much afraid that I shall not have the engine finished before you come ". The "Mr Penn" who is mentioned was the first John Penn, father of John Penn II, afterwards so well known as a marine engineer.

What Trevithick's ideas of the size of a screw propeller should be are to be inferred from this description:[1]

I have drawn the Cylind[er] 9 ft Diamr in which the screw is to work, which will be abt 8 ft 10 In Diamr; which have two turns in 4 ft long. Of course it will gain two feet forward to each turn, and each round to the steam arms, making the same to the screw, being on the first motion. At a rate of aboute 350 ft pr Sd on the end of the arm, will give the screw abt 10 miles forward pr. Hour.

Whether the screw propeller was actually tried on the Thames or not we cannot confidently say, but obviously there was a very limited time in which to do so before he sailed from England in October for South America.

Before closing the present chapter it is worth while to mention two minor matters that occurred during the period under review.

The first is the sound advice given by Trevithick in a favourable report which he made to certain London adventurers on the prospect of reworking Tregonick Mine. He concludes that it is[2]

well worth risking the sum of money proposed for that purpose, still

---

[1] *Enys Papers. Life*, i, 369. Letter to Giddy, dated Bridgnorth, 1815, May 12th.

[2] *Draft Letter Book.*

I wish each and every one of the Adventurers to look on it as a risk and an adventure as all mining concerns are. Therefore every one who may engage a share on my report ought to make up their minds to advance their money on a concern where I think the favourable prospects will warrant the prosecution of the mine but If it turns up a blank they must not sencure me for there are no certainty in Adventures.

The second matter is the fair dealing on Trevithick's part in his negotiations for the sale of a share in a mine, Wheal Francis, of which he appears, nominally at least, to have been the sole adventurer. Trevithick insisted that the prospective purchaser should satisfy himself by sending a competent miner to report upon the value of the mine. This the purchaser failed to do and Trevithick gave him the names of several competent miners: "Anyone but Captain Andrew Vivian". At length Samuel Grose acted, and from the Wheal Francis Cost Book we find that the transaction was carried through in October, 1816, shortly before Trevithick's departure for Peru.

PLATE XV. RICHARD TREVITHICK, AGED 45

From the oil painting by Linnell in the Science Museum, South Kensington

CHAPTER V

THE GREAT ADVENTURE

Silver mines in Peru—Arrival of Francisco Uvillé in England—Engines for
high altitudes—Takes shares in Cerro de Pasco Mines—Exports machinery
—Sails for Peru—Jealousy and intrigue—Prospecting for copper in Chile—
Revolution and civil war—Destruction of mines—Meets Simon Bolivar and
Lord Cochrane—Devotion of Mrs Trevithick—Visits Costa Rica—Journey
through forests to Caribbean Sea—Meets Robert Stephenson at Cartagena—
Returns to England.

WE have seen that Trevithick on his return to Cornwall
in 1810 resumed his former occupation of engineer in
the mines, and during the next six years had substituted his high-pressure boilers for the wagon boilers for use with
the larger mine pumping engines; had invented the plunger-pole
engine of which he built many during the period; had compounded
several pumping engines by the addition of a high-pressure
plunger cylinder; had invented the single-acting expansive
engine, which he applied to agricultural and other purposes;
and was experimenting with his recoil engine and screw propeller. He was now in his forty-second year—the very prime of
life—when there occurred an event which stirred his imagination to its utmost depths, and influenced the remainder of his
life.

The ancient silver mines in the Cerro de Pasco district of
Peru, situated at an elevation of over 14,000 feet above sea level
and about 160 miles from the capital city of Lima, had fallen into
a deplorable state, largely for the want of adequate means of un-
watering them. In 1811, a man of Swiss origin, named Francisco
Uvillé, had discussed with two prominent merchants of Lima,
Don Pedro Abadia and Don José Arismendi, the possibility of
accomplishing this by means of steam engines. Accordingly
in that year Uvillé was sent to England on a mission of enquiry.
Beyond this fact we know few details of his doings. He appears
to have got into touch with Boulton and Watt during this

visit and to have been informed by them that the low vacuum to
be attained in the attenuated atmosphere at the great height of
the mines would be insufficient to enable the low-pressure engine
to work to any advantage; further that it was impracticable to
construct engines in small enough pieces to be carried over the
only available road, a mere mule-track which rose to a height of
17,000 feet above the sea. While in London Uvillé saw, by a
mere chance, in the window of the shop of Wm. Rowley, 7,
Cleveland Street, Fitzroy Square, described in the Directory as
an "engine maker", a model of a steam engine. Enquiring
what it was, he was told that it represented the high-pressure
steam engine of Richard Trevithick. How the model got there
we are at a loss to imagine, unless Rowley had had some employment on the London locomotive, the site of the trial of which
was not far distant, and the model had been deposited with him.
Uvillé bought the model for twenty guineas and returned with it
to Lima; thence it was carried to Pasco, where it was found to
work satisfactorily.[1]

Uvillé now with his two merchant associates formed a company to drain the Pasco mines. An agreement was entered into
between them and signed on July 11th, 1812, the salient points
of which are as follows:

The Company to consist of the three contracting parties without admitting any other whatever.

The capital to be $40,000 (say £8000), of which Abadia and
Arismendi each held two-fifths, while Uvillé held the remaining
fifth. Uvillé was authorized to spend $30,000 (say £6000), the
estimated cost of two engines in England. If he could obtain
another engine on credit he was to purchase it on account of the
Company. Should the undertaking yield profit Uvillé was to be
credited with $2000 (say £400) for the value of the model.
Uvillé was also authorized to engage one or two English
workmen.

On September 26th, 1812, an agreement was signed by the
adventurers and the various miners interested. In this the
Company undertook to bring engines from England, take them

[1] *Life*, ii, 228. Letter, R. Edmonds to H. F. Boaze.

to the mines, and place them in Yauricocha and afterwards in Yanacancha, Caya Chica, Santa Rosa, and in the mining ridge of Colquijilca, and construct a pit for the collection of waters to a stated depth below the adit at Santa Rosa. The Company were to take in payment 15 per cent. in the case of Yanacancha and Yauricocha and 20 per cent. in the case of the other mines of the value of the ore raised. The Company undertook to commence to pump within eighteen months from the time of the signing of the contract and were granted the monopoly for nine years.[1]

Uvillé sailed once more on the long voyage round Cape Horn to seek in England the man who had invented the high-pressure engine but of whom he knew nothing else. He appears to have talked freely of his intentions, and bets were made among those on board as to the probability of his success. During the voyage he fell ill with brain fever and was landed at Jamaica, where he recruited, at last taking his passage for England in the Falmouth Packet "Fox", Captain Tilly. On board, still talking of the object of his voyage, he was informed by a fellow-passenger named Teague that nothing was easier, because the man he sought was his cousin, Richard Trevithick, who lived not far from Falmouth to which port the "Fox" was bound. There Uvillé landed on or about May 10th, 1813, after a journey that had lasted altogether six months. He was not yet wholly recovered from his illness, and was carried ashore to a boarding establishment on Falmouth Moor, now a private house. Notwithstanding that he was still confined to his bed, he wrote to Trevithick, who appears to have visited him at once. On May 20th Trevithick wrote to Rastrick:[2]

Yours of the 7th inst., I should have answered by return as requested; but an unexpected circumstance prevented my being at Swansea as early as proposed, which, as it happens, best suits your purpose as well as my own. I shall not be able to be there within twenty days from this time, of which I will give you timely notice. I hope before that time Mrs Rastrick will be safe out of the straw. I have been detained in consequence of a strange gentleman calling on

[1] *Life*, II, 222.
[2] *Life*, II, 195. Letter dated Camborne, 1813, May 20th.

me, who arrived at Falmouth about ten days since from Lima, in South America, for the sole purpose of taking out steam-engines, pumps and sundry other mining materials to the gold and silver mines of Mexico and Peru. He was recommended to me to furnish him with mining utensils and mining information. He was six months on his passage, which did not agree with his health, and has kept his bed ever since he came on shore; but is now much recovered, and hopes to be able to go down in the Cornish mines with me in a few days. I have already an order from him for six engines, which is but a very small part of what he wants. I am making drawings for you and intend to be with you as soon as they are finished. Money is very plentiful with him, and if you will engage to finish a certain quantity of work by a given time, you may have the money before you begin the job. The West India engine will suit his purpose. I shall have a great deal of business to do with you when we meet.

Trevithick's imagination was greatly fired by Uvillé's arrival, and there is little doubt also that his enthusiastic energy reacted upon Uvillé. The fact that in less than a fortnight after his first interview with the invalid on his sick-bed, Trevithick should have received an order from him for six engines, and that he was already making drawings of them, takes away our breath.

Two days after writing the above letter to Rastrick, Trevithick writes to him again, giving particulars of what was required, and the day after that to Pengilly of the Neath Abbey Ironworks in similar terms:

Below you have the particulars of what is wanted to compleat an engine and Pitwork, a 24 In. cylinder single 6 ft. Stroke a single Nozzle with two 5 In. valves, perpindicular pipe, piston and Cylinder bottom a 3 In. Safety-valve, fire Door, two small Y. Shafts and handles and a piston rod not turned and no cover to the Cylinder, a good strong winch sett in a brandice[1] frame such as is often used on quays or in quarrys and a force pump to feed the boiler, 25 fathoms of 12 Inch pumps a 12 Inch plunger 11 Inch working barrall with clack seat and windbore with Brass Bucketts and Clacks and abt. 10 fm of 3 In. pipes to carry water to and from the boiler is what I call a sett, six full setts of the same kind I want delivered in Cornwall as early as possible, a part of the Wrought Iron for the pitwork and boilers I have agreed with som smith in Cornwall, now I want your immidate answer saying

---

[1] 'Brandice' is a Cornish term for a three-legged frame.

part of these materials you will enter in to an engagement to deliver on Bristol Quay within four months from the time the drawings are sent you, I will lodge in a bank the money to pay for the whole of the Mat^{ls} and it shall be inserted into the same agreement that you shall draw every week for your money as you turn it oute at Bridge North, immid^{ly} on receipt of yours you will hear from me again.[1]

On June 2nd, Trevithick writes to Uvillé, who was still convalescing at Falmouth:

Yesterday I engaged a great many smiths and boiler-makers who set to work this morning. I have also engaged all the boiler plates in the County.... The master-smiths that I have engaged are the best in the kingdom. I have obligated them to put in the best quality of iron....Mr Teague is with me, and one other, assisting about the drawings.... As soon as it is convenient to you to arrange the payments I would thank you to inform me, because we find in practice that the best way to make a labouring machine turn quickly on its centres is to keep them well oiled.[2]

From Trevithick's reference to Teague as a man known to Uvillé, it seems very likely that he was the same person who had directed Uvillé to Camborne. We refer to him here, as although he appears in a very dubious light, we owe to him the introduction of an element of comedy into this narrative.

On June 8th, Trevithick writes to both Rastrick and to Price of the Neath Abbey concern, enclosing drawings of pumps and castings for boilers to enable them to make a start.

By June 11th, Uvillé had so far recovered his health as to have made the journey to London to put the financial side of the undertaking in train. On that date Trevithick writes to him to "Campbell and Co.'s Bank Buildings, London", saying that he was going to South Wales on the Lima business, then on to Bridgnorth there to meet Uvillé.[3]

Writing again on June 19th, he presses for "a few drops of that essential oil that you proposed sending me on your arrival in town".[4]

---

[1] *Draft Letter Book.* Letter to Rastrick, dated Camborne, 1813, May 23rd.
[2] *Life*, II, 199. Letter dated Camborne, 1813, June 2nd.
[3] *Life*, II, 201. Letter dated Camborne, near Truro, 1813, June 11th.
[4] *Life*, II, 202. Letter dated Camborne, 1813, June 19th.

Uvillé was now rapidly placing himself in a difficult position, for he had already far exceeded his instructions in ordering machinery, and was finding difficulty in raising money. We must recollect that he had been in London only a day or two, and we suspect that he was sufficiently a Spaniard by habit, to be quite unused to Trevithick's hurricane methods of doing business. By the 23rd no remittance had been received and Trevithick in a strong letter to Uvillé declined very naturally to proceed to South Wales and Shropshire on the business till money was sent,[1] and on July 4th he writes: "I intend to leve Truro for London on tuesday morning".[2]

We hear nothing further of the Neath Abbey Ironworks. Probably they would not accept the contract on the terms which Trevithick imposed as regards delivery. It is clear that he wished to divide the order between South Wales and Bridgnorth with the object of expediting completion. Eventually the whole of the engines and castings was made by Rastrick's firm and delay resulted.

By the 18th August Trevithick and Uvillé had returned to Cornwall, after having visited Bridgnorth together. Hazledine, Rastrick and Co. were evidently anxious to propitiate the foreign customer, as Trevithick's letter to them on that day indicates:[3]

I have your favor of the 10 Ins^t containing the invoice of sundrys from Bristol to Cornwall with pipes and Ale all of which being very necessary tools for engineers and miners, the Ale and pipes we are very thankfull for and shall not forget to remember each your health often over a glass of good Shrop Shire ale which will be chearfully joined by my rib as long as we find the[y] can all continue to run. Mr Uville remains at my house and Spends the greater part of his time in the mines which he is much pleased with, was underground at Wh^l Alfred last Monday. We arrived at Swansea the Saturday after we left Bridge North. I did not meet the gents in the Coal mine but they are now on the Spot. We Sail'd from Swansea on Monday

[1] *Life*, II, 202. Letter dated Camborne, 1813, June 19th.
[2] *Draft Letter Book.*
[3] *Draft Letter Book.* Letter to Hazledine, Rastrick & Co., dated Camborne, 1813, Aug. 18th.

Morning and arrived at Portreath on Wednesday morning. An American ship two days after took three Swansea coal ships for Cornwall. We saw a Ship of this description near Lundy Island ab^t 4 miles from us but did not suspect her at that time, however, a miss is as good as a mile. I was informed at Swansea that the Bristol Packett did not continue to sail, therefor I wish your wharfingers to Ship the Castings direct for Cornwall and If you have occasion to write them please to remind them of the necessity of getting these goods down. Mr Uville have not yet arranged any thing farther respecting the fourth engine but will write you in a few days on that Subject, therefore you are better to get finish'd as quick as possible the order given you, Long before which time I hope you will receive the order for the fourth engine. Do not neglect to write us often and let us know how you are getting on, Mr Uville and Mrs T. joins me in their best respects to all friends in and round the Bridge North Foundry.

This characteristic letter is a reminder that the second war with the United States of America was being waged at this time and that her cruisers were round our shores adding to the worries and hazards of the peaceful trader. On September 7th, Trevithick again writes to Hazledine, Rastrick and Co., ordering the pumping engine mentioned above and also "one Crushing apparatus compleat in addition to the former order", and states that Uvillé had received a letter from the Spanish Government saying that a line-of-battle ship would take the machinery to Peru, and indeed was detained for that purpose; he pressed them to push forward the work for shipment to Cadiz whence she was to sail.[1] Trevithick was energetically pushing the boiler and smith's work in Cornwall. Money was now coming in, but in small amounts at a time.

Writing again on September 22nd, he says that they will soon be at Bridgnorth, when "you will be very much annoyed with our company unless we fiend your assertions grounded on facts...for nothing short of a want of Cast Iron will confiend our friend in England one day after the end of this month".[2]

To another correspondent he says: "I shall be at Penryn with Mr Uville the latter end of next week to Ship the materials for

---

[1] *Draft Letter Book.* Letter dated Camborne, 1813, Sept. 7th.
[2] *Draft Letter Book.*

London. . . . I shall go to London in ab^t 10 or 12 days and will fiend oute the compleatest Dock and make a correct drawing of all the apparatus ".[1] This shows that he was busy in all directions.

He gives a sketch (see Fig. 29) in a letter to Bridgnorth, dated October 11th, pointing out a possible error in the arrangement of the winding engines whereby the flywheel would prevent access to the working cock, and expresses the fear that the various delays would cause them to be too late for the sailing of

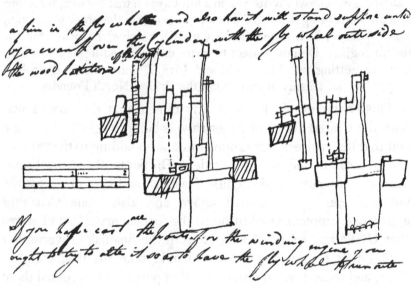

Fig. 29. Trevithick's winding engine for South America, 1813

the Spanish line-of-battle ship. He adds: "I fear that Mr Rastrick being so much from home will much impead our job".[2]

Rastrick was at this time engaged on the construction of Chepstow Bridge. Trevithick's fear proved well founded, and on December 28th he writes to Rastrick, saying that "a ship will sail for the South Sea fishery in about five weeks. . . .I shall be in Bridgnorth in about ten days and will remain untill the work is finished".[3]

[1] *Draft Letter Book.* Letter to Richard Symons, dated Camborne, 1813, Sept. 20th.

[2] *Draft Letter Book. Life,* II, 207.

[3] *Life,* II, 212. Letter dated The Plough Inn, Blackwall, 1813, Dec. 28th.

We have already referred to the Teague comedy and we must now briefly explain what it was. Among Trevithick's draft letters is one to "Mr John Teague, Penryn", repudiating a demand for payment of an account which he said he did not owe. On November 6th, writing to Edwards, a Truro lawyer, he explains the circumstance:[1]

This Day I rec^d a Summons from Lord De Dunstanville to appear at Camborne on Tuesday next to answer the Complaint of John Teague for wages due from me to him for his services aboute the Engines for South America. I never employ'd him and the onely thing he did for me was making a small and very imperfect coppay of a drawing for the Smiths work which was from his own application to larn to draw which was not 6 hours service to me. From what I have been informed I believe Teague have reported a great many false things with a view to doing mistive and also that I am aboute to leve my Country to the great injury of the Cornish Copper mines. I wish your advice in what I am to do on tuesday and If you think a letter or the attendance of a person from you should be necessary, you will be so good as to take steps accordingally and write me in time. If the receipt given by him for £15 to Mr Uville will be of service please to enclose it.

Uvillé had been arrested at Teague's instance and the case was ultimately brought into court at Launceston Assize. The following is the sequel, told in Trevithick's words in a letter to Uvillé:[2]

Teague's Attorney informed Mr Edwards that he was present at Penryn when you informed him you acknolidged Teague's debt and that he would give such evidence at the Assize, in consiquence of which Mr Edwards sent a supena to both me and my whife to give evidence in court that Teague acknolidged when you paid him the £15 that he had no farther demand on you. We both attended the Assize and when the rascal of Attorney saw us both in court he drop^d the case. Mr Edwards said that as he new him to be in the habit of making false oaths that he would have done the same hear, if it had not been for two oaths direct against him. Teague had another sute there for Debt which he also lost which was equally as rascally as that of yours, Smith Truman of Penryn was bail for him and Teague have

[1] *Draft Letter Book*.
[2] *Draft Letter Book*. Letter to Uvillé, dated Camborne, 1814, March 31st.

absconded from his bail and the poor old Smith will have all his furniture taken to pay Teague's debt and Cost which will be above £80. A more compleat fool and rogue than Teague was never. Jane return'd in big Spirits from the Assize and begs her best respect and services to you in every occasion when call'd for. Your Cornish Shipmates are anxiously waiting your call.

Meanwhile the Bridgnorth firm was pressing Uvillé for money, eliciting caustic remarks from Trevithick. To get over his difficulties Uvillé now agreed to admit Trevithick as a partner in the Company, and a document to this effect was drawn up and signed by them both on January 8th, 1814.[1]

Under this agreement the capital of the Company was increased, Trevithick finding about £3000 and being admitted as a partner to the extent of one-fifth of the whole capital. Uvillé undertook that this alteration in the constitution of the Company should be ratified by Abadia and Arismendi on his, Uvillé's, arrival at Lima. As soon as this arrangement was completed Trevithick proceeded to sell part of his share to raise money to liquidate the unpaid balance of the debt incurred in buying the engines, machinery, tools, etc. Page and Day, London lawyers, who were acting for Uvillé, as well as John Mills of Limehouse, and James Smith of Greenwich, men who, as we have seen in the preceding chapter, had both done business with Trevithick, appear to have taken shares. Page was selling shares and Trevithick was urging him on, just as he was everyone concerned. Writing to Page on March 30th, 1814, he says:

I rec'd your favor of the 12 Inst in which you said that next week I will write you, but not yet receiving it I conclude that London time is above double that of the Cornish when you fiend that your next week is up I doubt not but that you will fulfill your promise.

He then gives an account of Teague's lawsuit at Launceston and ends up by kindly enquiring after Mr Day's health. There appears to have been a little sparring between Trevithick and Page; we shall meet with this gentleman later.

A letter written to Hazledine, Rastrick and Co. in April, asking for their final account, is quite typical of Trevithick's

---

[1] *Life*, II, 218.

correspondence. He had written them many times during the
progress of the work, scolding them with biting sarcasm. He
now addresses them as "Most Worthy Gentlemen". The end of
the job was in sight and his ire had evaporated.[1]

He had engaged two Cornishmen to go to Peru with Uvillé
to erect the machinery. One was William Bull of Chacewater—
no relation as far as we know of his early associate—and the
other Thomas Trevarthen of Crowan. Of Bull, Trevithick
writes that he was only capable of acting as assistant to an
engineer, while Trevarthen was a capable pit-man and could
help in erection. Trevithick urged Uvillé to take with him in
addition a capable engineer. It seems that Henry Vivian,
brother of Andrew Vivian, brother-in-law of Trevithick, was
anxious to go, and had already approached Uvillé with this idea.
Trevithick writes of him that he is a suitable man except for
"that one failing of making too free with an evening glass, which
you were not unacquainted with while in Cornwall at Dolcoath
Mine", but that he, Trevithick, would take no active part in the
business, lest he should be blamed by the family if any accident
should happen to Vivian.

At last all the machinery and apparatus was shipped on board
the South Sea Whaler "Wildman", John Leith, Master, and
Uvillé, Bull, Trevarthen and Vivian sailed in her from Portsmouth for Callao, the port of Lima, on September 1st, 1814,
fifteen months after Uvillé had arrived in England.[2] As to
details of the machinery sent out we learn something from
Trevithick's letters and account books, as well as from a summary of the invoice of all that was shipped in the "Wildman".
There were four pumping engines with pit-work, four winding
engines, a portable rolling-mill engine, two crushing mills and
four extra boilers; besides spare parts and tools for miners,
blacksmiths and carpenters. Of the engines we find information
in Trevithick's letters to Rastrick and Pengilly (see p. 162 *ante*).

The engine for the rolling mill for the Mint at Lima had been,

---

[1] *Draft Letter Book.* Letter to Hazledine, Rastrick & Co., dated Camborne, 1814, April 8th.

[2] *Life,* II, 217.

as already mentioned, intended originally for R. W. Pickwood of St Kitt's in the West Indies, and was no doubt of the type Trevithick was then building for agricultural purposes. It will thus be seen that all the engines were of the single-acting open-topped cylinder expansive design; in fact, except for one or two engines for the Cornish mines, all the engines he built between 1810 and 1816, when he went to Peru, were single-acting. All the boilers appear to have been of the "Cornish" type, for we have evidence that the one supplied for the Mint was similar. We have some particulars, supplied sixty years later, of the engines (see p. 188) and they confirm the above statements. The summary of the invoice given in the *Life* gives the total amount, including freight and insurance, as £16,152. 1s. 1d. Trevithick's books show a rough account of material supplied to Uvillé amounting to £10,886. 12s. 6½d., but this appears to be incomplete. Both sides of the account are incorrectly added, and we are unable to say what was the amount of freight, insurance and other charges. Whatever the sum, the fact was that Uvillé had spent a sum greatly in excess of that authorized by his agreement with Abadia and Arismendi.[1]

Uvillé and his party arrived at Callao in January, 1815, as appears from a letter to Giddy:[2]

Aboute a fortnight since, I rec^d Letters from Lima, and also letters to the friends of the men who sail'd with the engines. They arrived the 29 Jany after a very good passage and withoute one hour's sickness. Both theirs and my agreements was immid'ly ratifyde and they are in big spirits. The ship finish'd diss^g the 11 Feby which was the day these letters sail'd from Lima with 12,000 Dollars for me which is all arrived save. I shall make another fit oute for them immidly. I expect that all the engines will be at work before the end of October. Half of them must be at work before this time. The next day after their letters sail'd for Europe, they intended to go back to the mines.

A further report, in a letter of April 19th to Trevithick, was also good; one engine was then at work and worked well: "the

[1] *Life*, ii, 220.
[2] *Enys Papers*. Letter, Trevithick to Giddy, Camborne, 1815, July 8th.

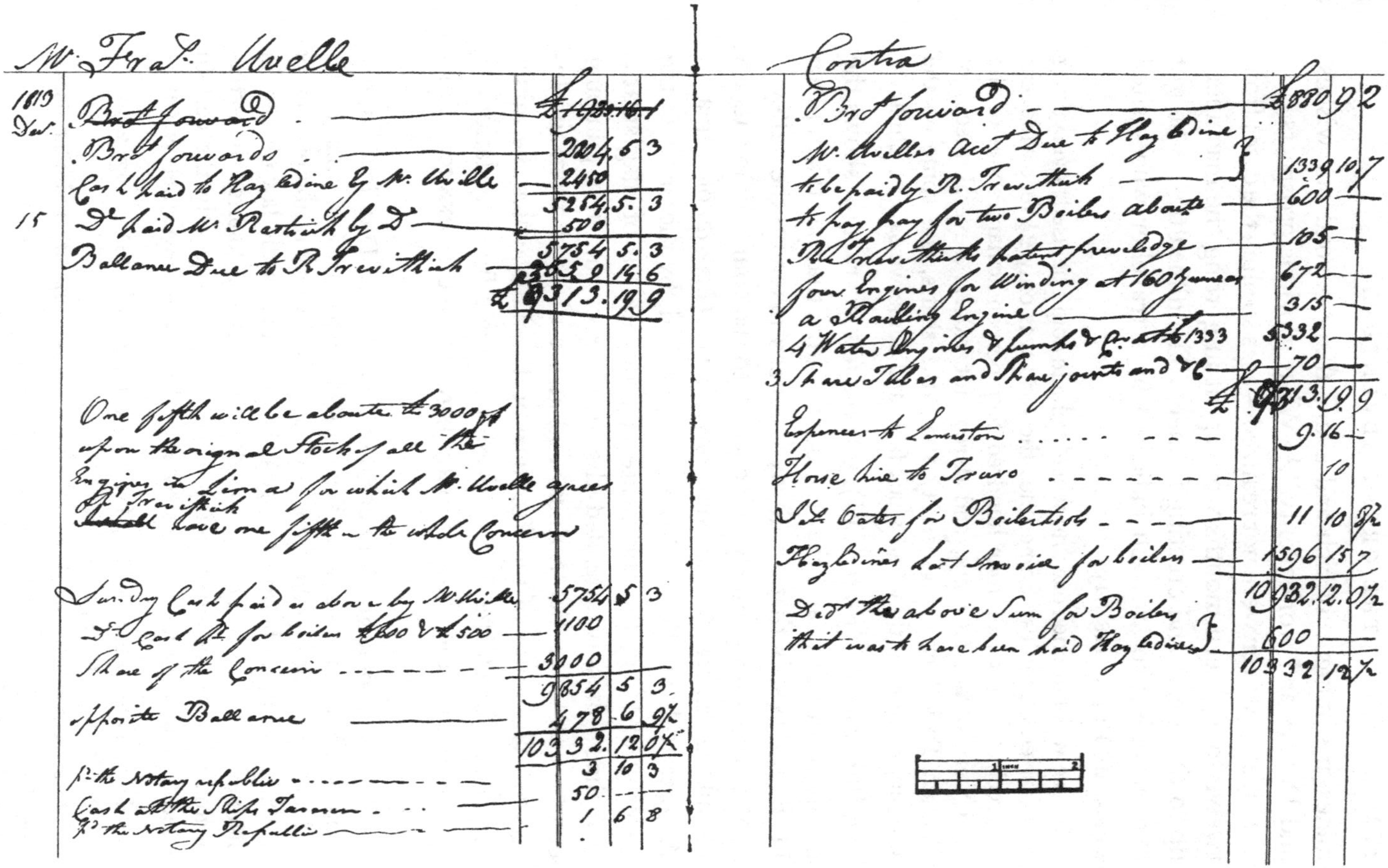

Fig. 30. Ledger account with Uvillé. From Trevithick's Account Book

mines are immensely rich and every thing beyond our wishes, and Dollars and work plenty".[1] What glorious prospects!

Hopes however were not realized—as is so often the case; from twelve to eighteen months were spent in getting the machinery up to the mines and erecting it there. The difficulties to be got over in doing so are well indicated by a traveller who visited Pasco in 1850. He says:[2]

There was but one road; no wheel vehicle could be used; everything was carried on mules. Sometimes the road was only $2\frac{1}{2}$ feet wide, cut in precipices three or four hundred feet perpendicular; some of the men were afraid to walk, and dared not ride.... several pieces of Captain Trevithick's engines lay about the shafts, and some on the way up, as though they had stuck fast, and some we saw at Lima.

Our illustration[3] (Plate XVI), taken a few years ago, shows the nature of the track and that the only change that has taken place in the interval is that a railway has been taken up the valley.

We have documentary evidence as to the course of events up to July 27th, 1816, for the *Government Gazette* of Lima of August 10th and September 25th published, by the order of the Viceroy of Peru, the certificate of a deputation of magistrates who had officially attended the starting of the first pumping engine erected at Santa Rosa, and also a dispatch from the In-tendant-Governor of the Province of Tarma in which the mines are situated. Of the first of these we give a full translation, as exemplifying the flowery language of these exalted func-tionaries:[4]

We, Don Domingo Gonzales de Castañeda and Don José Lago y Lemus, Commissaries and Territorial Magistrates in this Royal Mineral Territory, and deputed by the United Corporation of Miners in this district, do hereby certify judicially, and as the law directs, in manner following:—Though this deputation never doubted the extra-ordinary power of steam compressed, and consequently the certain operation of engines worked by its influence it nevertheless enter-

[1] *Draft Letter Book.* Letter, Trevithick to Rastrick, 1815, Aug. 24th.
[2] Mr Rowe, *Life*, II, 232.
[3] We are indebted to Mr C. O. Becker for this and other similar photo-graphs.
[4] *Life*, II, 233.

PLATE XVI. MULE PATH FROM LIMA TO CERRO DE PASCO
THROUGH THE ANDES, PRESENT DAY

Photo by C. O. Becker, Esq.

tained some fears respecting the perfect organization of all the mechanical powers of the machines. This uncertainty, rather than any doubt, has been completely dissipated by our personal attendance this day to witness the draining of the first pit, situated in Santa Rosa. The few instants employed in the same produce a full conviction that a general drainage of the mines will take place, and that their metals will be extracted with the greatest facility from their utmost profundity; as also that the skill of the company's partners and agents will easily overcome whatsoever difficulties nature may oppose, until they shall have completed all the perpendiculars and levels, and consequently that the meritorious undertakers who have risked their property in the enterprise will be rewarded with riches. We and the whole Corporation of Miners would do but little were we to erect them a monument, which would transmit down to the remotest posterity the remembrance of an undertaking of such magnitude and heroism; but for the present we will congratulate ourselves that our labours, co-operation and fidelity, keeping pace in perfect harmony with the exertions of the agents, the company may thus attain the full completion of their utmost wishes, extracting from the bowels of these prolific mountains, not the riches of Amilcar's inexhaustible wells, not the treasures of the boasted Potosi in its happiest days, but a torrent of silver, which will fill all surrounding nations with admiration, will give energy to commerce, prosperity to the Viceroyalty and to the Peninsula, and fill the royal treasury of our beloved sovereign. Thus certifies this Magisterial Deputation of Yauricocha, the 27th of July, 1816.

Domingo Gonzales de Castañeda
José Lago y Lemus.

The Governor's covering dispatch gives the further information that the first winding engine had also been put down at Santa Rosa. From the Viceroy's reply to these grandiloquent lucubrations, it appears that Bull had the chief credit for the erection. Meanwhile the engine at the Mint had also been erected. But under date June 4th, Uvillé writes to Trevithick saying that the Mint engine is not working satisfactorily, to which he replies:[1]

I yesterday rec^d your favor of the 4 June last and am sorry to fiend that the Mint engine do not go according to expectation. The Tube is not large enough to Burn wood, as I allways said therefore I expected you

---

[1] *Draft Letter Book.* Letter dated Penzance, 1815, Dec. 9th.

to put the fire under the boiler and return back through the tube and
then round the oute sides, then you would get steam enough to work
30 Pounds to the inch 30 Strokes per m^t [for] which the boiler is Large
and strong enough. Since you left England I work^d a boiler 2 ft Diam.
$\frac{3}{8}$ thick with 300 Pounds to the inch which it stood very well and If the
cock of the Mint engine did not open and shut in the proper time you
might put an iron rod to work it like those you saw of mine in London;
then I am shure of it doing everything you could wish it. I have not
yet put to work the great engine I mentioned to you by way of
Jamaica in my letter of last August but hope to get finish^d before the
end of this month it will do wonders. As soon as it is finish^d I intend
to see Lima and hope to be with you in a short time. I have made enquiry after South Sea Walers but cannot find any one that will undertake to put me on shore at Lima, fearing that the[y] should not tuch
there in the Course of their voyage; therefore I have determined to
come over land. I hope that before my arrival that all the engines will
be at work and the Dollars plenty. I fear nothing if political afairs do
not interfear with us. The newspapers say a great deal aboute the
State of Peru being disturbed by factions contending for independence.
I think it ought not to injure as you have not a wish or parshallity
either on the one side or the other. I have removed my family to
Penzance for the Advantage of being near the school for the children
to which place in future I would thank you to direct my letters,
Mrs Trevithick joins in best respects to you and is glad to hear you
are well. I intend to Bring oute with me for you a very good working
moddel of this new engine. The Cornish mines are very poor, the
price of Tin and Copper very low, and a great many miners in want of
employ. I wish a thousand or two was at Pasco. Mr Dubois's send
me your letter Dated last April; the[y] are all well and have wrote you
several times; the remittances both to them and me we have rec^d. The
Ship that transported Buonaparte have returned and left him at St
Helena. Tho' we are at Peace with all the world, yet extraim distress
prevails, no trade no money nor no faith and nothing but a general
ruin expected under peace.

Trevithick also wrote to his men in Peru about the Mint
engine, saying:[1]

I wish you had put the fire under the boiler and through the tube,
as I desired you to do, in the usual way of the old long boilers, then
you might have made your fire-place as large as you pleased, which

[1] *Life*, II, 227.

would have answered the purpose, and have worked with wood as well as with coal, and have answered every expectation. I always told you that the fire-place *in the boiler* was large enough for coal, but not for wood, and desired you to put it under it.

He again refers to his intention, since he could not succeed in finding a South Sea Whaler which would undertake to land him at Callao, to sail for Buenos Aires and cross the continent to the Pacific coast.

The men who went out with Uvillé were capable enough when under skilled supervision in their native mines in Cornwall, but had no experience in pioneer work of this kind, nor could they be relied upon to appreciate Trevithick's forethought in allowing for eventualities which might have to be faced. Seeing this, he became more anxious than ever to be on the spot himself. He at last secured a passage in the South Sea Whaler "Asp", Kenny, Master, which was calling at Callao, but it was not till October 20th, 1816, that he walked down to the harbour side at Penzance with his young son (then a boy of four years of age) and future biographer on his shoulder, and embarked upon the long voyage from which he was not destined to return until eleven adventurous years had elapsed. With him sailed Richard Page and James Sanders or Saunders, a boiler-maker, probably a son of the man of the same name who had for a long time worked for Trevithick. A word of explanation is necessary here with regard to Page. We learn from John Miers,[1] who was acquainted with Abadia and his family and was in Lima at the end of the year 1823, that the London shareholders in the company sent out an agent to look after their interests. We learn also from Richard Edmonds[2] of "a gent who had lately arrived from England". It will be remembered that Page was the London lawyer who had acted for Uvillé and Trevithick in the sale of shares, and there is little doubt that he was the man referred to. Page and Trevithick had apparently not always been friendly, and the latter had doubts as to the advisability of encouraging Page to go with the party to Lima. In this he was overruled by his wife, who was

[1] Miers, *Travels in Chile and La Plata*, 1826, ii, 441.
[2] *Life*, ii, 246.

naturally pleased at the idea of her husband having a companion on the long voyage he was about to undertake.

In Trevithick's draft letter books are copies of letters received by his wife and others after her husband's departure. They were evidently copied into the book as a record, in a youthful hand, most likely that of one of Mrs Trevithick's children. The first news we have of Trevithick in South America is contained in one of these dated "Callao, Feb. 18, 1817", and commences "My dear Sister". It was written the day after the "Asp" had dropped anchor in Callao Bay after a favourable passage. The party were awaiting an order from the Viceroy to permit them to visit Lima. The letter was written so as to catch a ship just sailing from the Bay, and at that time the party had not set foot ashore since leaving Penzance. The writer describes the view from the ship's deck, whence "the Stupendous Andes, far higher than the clouds, seem like the boundary of the World". Arismendi had been on board and Abadia was expected: "all were very glad to see Trevithick". There can be little doubt that this letter was written by Page, because there was no one else on board who could have written in this strain. The letter goes on to say:

I find however, that the expense of the engines have been very great and that it has been necessary to take in other Partners but as yet have none of the Particulars. My next which will be sent off in a month's time will tell you all about it...two of the engines are at work...and then I expect to have a good deal of information respecting the situation and Prospects of the Concern. Give my love to Mr Smith.

Miers states that[1]

upon his [i.e. Trevithick's] arrival in Peru a new agreement was made, in conjunction with the richest and most influential persons in the country, by which the concern was to be extended on a grand scale; an immense quantity of machinery of various kinds was procured from England, which is now lying useless in the cellars of the Mint at Lima where I saw it. It consisted of immense trapiches [i.e. sugar mills], grinding and amalgamating mills of cast iron, a series of rolling and laminating apparatus, all intended to be worked by the power

[1] Miers, *loc. cit.* II, 438.

of two steam engines which accompanied them, together with much furnace work for the refining and alloying of the silver.

We see that both Miers and the writer of the letter given above, who must have been Page, refer to this expansion of the Company's operations as happening about the time of Trevithick's arrival in Peru; the only mention we have of a further quantity of machinery being sent out from England is contained in his letter of July 8th, 1815 (p. 170) above, where he says: "I shall make another fit oute for them immidly". Soon after his arrival Trevithick wrote to James Smith of Greenwich:[1]

We arrived here last Saturday in good health. The (our) Mint is at work, and coined five millions last year, and in their way of working does very well; but I trust to make it coin thirty millions per year.

Two engines are drawing water, and two drawing ore, at the mines, but in an imperfect state. If I had not arrived, it must have all fallen to the ground, both in their mining and in their engines. I expect we shall go to the mines in about ten days, from where I will write to you every particular.

There are still two engines to put up for lifting water, and two for winding ore, and those at work to be put to rights. They are raising ores from one mine which is immensely rich, and from what I can learn, a much greater quantity will be got up, when the whole are at work, than these people have any idea of. Several other mines will also be set to work by engines that we shall make here. We have been received with every mark of respect, and both Government and the public are in high spirits on account of our arrival, from which they expect much good to result.

Mr Vivian died the 19th of May. I believe that too much drink was the cause of it. Uville, I think, wished him gone, and was in great hope that I should not arrive. His conduct has thrown down his power very much, which he never can again recover.

They all say that the whole concern shall be put entirely under my management, and every obstacle shall be removed out of my road. Unless this is done, I shall soon be with you in England. I am very sorry that I did not embark with the first cargo, which would have made a million difference to the company. The first engine was put to work about three months since, the other about two months; but they are as much at a loss in their mining as in their engineering. The Mint

---

[1] *Life*, ii, 240, date not stated.

is the property of our company, and Government pays us for coining, which gives us an immense income; the particulars of which, and the shares in the mines, I have not yet gone into. I shall be short in this letter, because I know but little as yet, and that little I expect Mr Page will inform you. A full account you shall have by the next ship, which I expect will sail in three weeks. This letter goes by a Spanish ship that will sail this afternoon for Cadiz. My respects, and good wishes to your family and to Mr Day, and hope this will find you all as hearty as we are.

Mr Page would not depart this life under the line, as he promised when at Penzance; but, on the contrary, has a nose as red as a cherry, and his face very little short of it. His health and spirits far exceed what they were in England. I am glad to have such a companion. With...think he will have no reason to repent....He will get a command at Pasco...such as his ingenuity may find out, when on the spot; whether as a miner or an engineer I cannot say, but time will show.

If you have not insured my life I would thank you to do it now, if you can on reasonable terms. I do not wish them to take the risk of the seas in the policy, because the voyage here is over, and on my return I hope I shall not want it, therefore it must be for two years in the country. I will get a certificate of my health, if they wish it, from the most respectable inhabitants, and also from the Vice-king, if they wish it. The policy may be drawn accordingly.

Be so good as to write me often, with all the news you can collect. If you wish your dividends in this company to be applied to further advantage in any new mines I may engage in, in preference to having it sent to England, I will, as the dividends are made, do everything in my power to improve the talent. On this subject I must have your answer before I can make any new arrangement under this head. I will thank you to send a copy of Mr Page's letter to my wife; I mean such parts of it as belong to the business; there may be some things that I have forgotten to mention.

A letter of Trevithick's written about this time contains an amusing account of his doings:[1]

There are also nunneries beyond number, and in those places no male is ever suffered to put his foot. Through one of the most noted runs a watercourse, which works the Mint; and Mr Abadia has repeatedly made all the interest he could to be admitted, for the purpose

[1] *Life*, ii, 243, also without date being given.

of inspecting it, but could never get a grant. The Mint belongs to our engine concern, and now coins about five millions per year. We have a contract from Government for making all the coin, both gold and silver, which gives an immense profit; and as there must now be coined six times as much as before, I must build new water-wheels to work the rolls which we took with us from England. It was on this account that I wished to examine the watercourse for this purpose, without the knowledge of Mr Abadia or anyone but Mr Page and the interpreter, who always attends me. I walked up and knocked, in my blunt way, at the nunnery court door, *without knowing there were any objections to admit men*; it was opened by a female slave, to whom the interpreter told my name and business. Very shortly three old abbesses made their appearance, who said I could not be admitted. I told them I came from England, for the purpose of making an addition to the Mint, and could not do it without measuring the watercourse; upon which a council was held amongst them; very soon we were ordered to walk in, and all further nunnery nonsense was done away. We were taken round the building and were shown their chapel and other places without reserve.

Uville knew nothing about the practical part of the engines, and Bull very little, therefore you may judge what a wretched state this great undertaking was in before my arrival; no one put any confidence in it, and believed it was all lost, together with five hundred thousand dollars that had been expended on it. The Lord Warden was sent from Pasco to offer me protection and to welcome me to the mines. They have a court over the mines and miners, the same as the Vice-Warden's Court in England, only much more respected and powerful. The Viceroy sent orders to the military at Pasco to attend to my call, and told me he would send whatever troops I wished with me. The Spanish Government and the Vice-king since my arrival are quite satisfied that the mines will now be fully carried into effect, and will do everything in their power to assist me. As soon as the news of our arrival had reached Pasco, the bells rang, and they were all alive down to the lowest labouring miner, and several of the most noted men of property have arrived here—150 miles. On this occasion the Lord Warden has proposed erecting my statue in silver. On my arrival Mr Uville wrote me a letter from Pasco, expressing the great pleasure he had in hearing of my arrival, and at the same time he wrote to Mr Abadia that he thought Heaven had sent me to them for the good of the mines. The water in the mines is from four to five strokes per minute.

Tell members of the Geological Society that Mr Abadia is making out a very good collection of specimens for them, which will be sent by the first opportunity; and soon after I arrive at Pasco I will write them very fully.

Trevithick was soon at Pasco, where he got the two pumping and winding engines in good order, and everything pointed to the future success of the enterprise. Uvillé, however, finding, as he had anticipated, that he was no longer regarded as the man to whom this success was due, did all he could to oppose Trevithick. In this he was supported by a clique among the shareholders, if indeed the opposition did not originate with them. From Miers's account, Trevithick appears to have been very badly used. They thought they could do without him. Obstacles were put in his way and they wished to get rid of him. The shareholders in London sent out an agent to look after their interests. Much larger sums were found to be necessary. Complaints as to delays and expenses were made on all sides and difficulties attributed to mismanagement.[1]

The greatest share of opprobrium unjustly fell upon Trevethick who being a man of great inventive genius and restless activity, was at length completely disgusted and retired from the undertaking. He left Pasco, although Abadia offered him 8,000 dollars per annum, together with all his expences, if he would continue to superintend the works.

There is considerable evidence that Page was the man who took a leading part in this opposition, if indeed he was not its chief instigator. Among the letters, referred to above in the draft letter books, of later date than Trevithick's departure from England is one without name, address, or date. The letter bears internal evidence of having been written by his wife, and of having been addressed to some of the London shareholders who had apparently criticized severely her husband's actions. She wrote:[2]

It gives me pleasure to hear of Mr Trevithick's return to the Pasco Mining Company, and have not the smallest doubt of a mutual and

[1] Miers, *loc. cit.* II, 440.
[2] *Draft Letter Book.*

lasting understanding between him and Mr Abadia provided Page on his arrival does not interfere. For some time previous to Mr Trevithick's sailing, he debated whether it was best to take Mr P. with him but I hoped he would have proved an agreeable and useful companion, and my husband against his better judgment consented to my request. Now however, I have no doubt of his intention before he left England to become a secret mover of some vile plot, having in consequence of a quarrel threatened that if ever he reached Lima, things should not be as they would have been, the meaning of which threat was not understood until he had put his plans into execution. Had Mr Trevithick read his agreement with him he would have suspected his intentions. I was at this time confined to my room, and having after my husband's departure inquired for different papers, discovered that Mr Smith had taken them to London with him, one of which was Mr Page's agreement. This was not returned until Mr Smith had received Mr Page's first letter from Lima; it was then conveyed through Mr Marshall and I immediately sent a copy of it to Mr Trevithick. When they arrived at Lima it appeared to Page that the best method of carrying his plans into execution would be to inflame Mr Uville's ambition and rouse his jealousy of Mr T. You must also be aware that this was the side to attack him. When this was effected the rest was easily accomplished, being already in Uville's interest and Sanders weak and easily frightened into it. Captain Hodge informed me that when the news of Mr T.'s arrival reached Pasco, Uville set out with a party intending to intercept him in his progress to the mines; in this however he was disappointed, and having reached Lima, he was obliged to receive him as others did, and this appears to be confirmed by a series of letters from Uville, Bull and Sanders at Pasco to Page at Lima, which were brought with him to England and sent to Cornwall for my perusal; these doubtless you have seen. Thus had the friends time to consult and consider such means as were put into immediate execution on their reaching the mines. Under such circumstances could your letter of the 16th of October in which you speak of the haughty imperious and brutal conduct of my husband fail to arouse indignation? Even the money which you were given to understand he was about to undertake some new project of his own account, Mr Abadia in his letter to Mr Tyack allowed he had never received. I am not informed of what Mr Tyack wrote to you, but I should suppose he proved Page's falsehood by asserting the boiler plates and iron to be his uncle's not his own as Mr P. affirmed. I trust I am fully sensible of your kindness and gladly avail myself of the opportunity you have afforded me in writing to my husband.

As a result of all these petty jealousies, Trevithick, in indignation at his treatment, had, somewhere about 1817, left the
mines. The "new project" that he was "about to undertake"
"of his own account" was to prospect for minerals along the line
of the Sierra. We cannot do better than quote a memorandum or
letter without date or address, written by Trevithick, probably
after his return to England, at the time he was attempting to
float the Caxatamba Mine for which he refused £8000. The
document is valuable because it is the only account we have from
Trevithick himself. It is as follows:[1]

In 1814 an arrangement was made between the miners of Peru and
myself for furnishing them with nine steam-engines and a mint, to be
executed in England and erected in the mines of Pasco; and in October,
1816, I sailed from England for that country, for the express purpose
of taking the management of those mines and erecting the machinery,
being myself a large proprietor of the same. The Government of
Peru was at that time subject to old Spain, under the immediate
superintendence of a Viceroy. The machinery having been erected, and
its sufficiency for the intended purpose of draining the mines having
been proved to the satisfaction of all parties, there was granted to me
a special passport by the Viceroy, for the purpose of travelling through
the country to inspect the general mining system, and to make the
native miners acquainted with the English modes of working. In
return for which Government conceded to me the privilege of taking
possession for my own benefit and account of such mining spots as were
not previously engaged. In this way I travelled through many of the
mining districts, and although I met with several unoccupied spots
which would have paid well for working, yet, being a considerable
distance inland, and requiring more capital to do them justice than I
could then advance, I abandoned for the time all ideas of undertaking
them.

To this, indeed, there was but one exception, and that was a copper
and silver mine, the ores of which are uniformly united, in the province
of Caxatambo.

When the patriots arrived in Peru, the mine was deserted by all the
labourers, in order to avoid being forced into the army. In this state it
remained for a considerable time; but on the Spaniards retreating into
the interior, I recommenced working; and to secure my right to this
mine under the new Government I at the same time transmitted a

[1] *Life,* II, 251.

memorial and petition to the established authorities, accompanied by
a plan and description of the mine, the result of which was the formal
grant, as exhibited in the Spanish document now in your possession.
It was not my good fortune to be allowed to follow up my plans, which
almost warranted a certainty of success. I had scarcely commenced a
second time when the Spaniards returned, and everyone again was
obliged to fly. The country, as is well known, continued for a long
time in a most distracted state, and I was ultimately compelled to quit
that part of Peru, robbed of all my money, leaving everything behind
me, miners' tools and about 5000*l.* worth of ores on the spot ready to
be carried to the shipping port. Numerous as my misfortunes had
been in Peru, and heavy as my disappointments, I felt none so sensibly
as this, because it was an enterprise entirely of my own creation, and
so open to view that I was enabled to calculate at a certainty the
immense value contained within the external circle where the copper
vein made its appearance in the cap of the mountain, and to be obtained
without risk or capital. However, revolution followed revolution, and
the war appeared to me to be interminable. Even Bolivar's arrival at
Lima made it still worse, for he forced me into the army, with my
property, which is not paid to this day, to the amount of $20,000;
and at his urgent solicitations, disgusted as I was with what I had seen
and suffered in Peru, I determined on quitting it for a time at least,
and on visiting Colombia. Being at Guayaquil I first heard the name
of Costa Rica and its recently-discovered mines, and having no doubt
of the authenticity of my information, I immediately proceeded thither
instead of going to Bogota to carry Bolivar's orders into execution,
not having been paid. This short digression you will excuse, as it
points to the causes of my separation from a property of so much
value, as I consider the mine of * * * * * * * *.

The document then goes on to state the low cost of labour,
living and transport; the ore was to be sent to England to be
smelted, although later it was proposed to smelt on the spot.
Apparently about 300 tons of ore had been worked, and a
vessel chartered to take it. Commenting on this document
Francis Trevithick says:[1]

He often spoke of his discovery and working of the great vein of copper
ore in Caxatambo, estimated to contain copper worth twelve millions
sterling, the working of which was prevented by the frequent revolu-
tions and unsettled government of the country; and of residing for

[1] *Life*, ii, 254.

months with Bolivar, at that time the Republican Governor of Peru.

Bolivar's cavalry were short of fire-arms. Trevithick invented and made a carbine with a short barrel of large bore, having a hollow framework stock. The whole was cast of brass, stock and barrel in one piece, with the necessary recess for the lock; the bullet was a flat piece of lead, cut into four quarters, held in their places in a cartridge until fired, when they spread, inflicting jagged wounds. He was obliged to serve in the army, and to prove the efficiency of his own gun. He was never a good shot, nor particularly fond of shooting; and, after a long time, Bolivar allowed him to return to his engineering and mining. Scarcely had he got to work again when the Royal Spanish troops, getting the best of it, overran the mines, and drove Trevithick away penniless, leaving 5000*l.* worth of ore behind him ready for sale.

The 300 tons of ore, valued at 24,000*l.*, never reached England; and the writer, who was to have returned to Peru in the ship that had been engaged to convey it, lost the chance of being a youthful traveller in foreign lands.

It is to this period, probably, that Miers refers when he says:[1]

He...entered into speculations with some of the miners at Conchucos, for whom he constructed grinding mills and furnaces with the view to substitute the process of smelting for that of amalgamation, in silver ores, in which vain pursuit he became a considerable loser.

An incident mentioned by Stevenson refers to the same adventure, although the date he gives is probably two years too early:[2]

In the year 1817, two Englishmen, sent from Pasco by Mr Trevithick, who afterwards followed with the intention of working some of the silver mines in Conchucos, were murdered by their guides at a place called Palo seco. This horrid act was perpetrated by crushing their heads with two large stones, as they lay asleep on the ground; the murderers were men who had come with them from Pasco.

In reference to this incident, we have a note from the journal of Captain Hodge, a Cornish miner at that time in Peru:[3]

The first time they met was at Lima, on the 26th of April, 1819, at Dr Thorne's; your father had just come down from the Cerro de Pasco mines. On the 28th May following I find my father witnessed the

---

[1] Miers, *loc. cit.* II, 441.

[2] W. B. Stevenson, *Twenty years' residence in South America*, 1829, II, 61.

[3] Given to Francis Trevithick by Hodge's son Charles, *Life*, II, 246.

hanging of three men for killing two of your father's men, named Judson and Watson.

In August, 1818, Uvillé died. Bull had died some months previously. All those who had sailed in the "Wildman", possibly Trevarthen too, of whom we hear nothing further, were now dead, and we find that Trevithick, soon after Uvillé's death, was in full control of the Cerro de Pasco Mines.

By this time the unrest which had for many years been seething in Spain's South American possessions was rapidly reaching open manifestation in Peru. Insurrections had broken out in various places years before Trevithick's arrival, and as early as 1810 the most important of these had resulted in the triumphal entry of Simon Bolivar into his native city of Caracas, in Venezuela, where the title of "El Libertador" was bestowed upon him. Throughout the varying fortunes of the opposing factions, Lima, the stronghold of Spain's military strength in South America, had remained comparatively tranquil, but in 1812 Lima and Mexico were the only capital cities still under the rule of the ancient authority. In February, 1817, shortly after Trevithick's arrival in Peru, a Spanish army was totally defeated by the insurgents at Chacabuco in Chile, followed in the next year by a counter-victory for Spain. Perhaps the most notable event of the war of secession occurred in February, 1819, when a Chilian squadron appeared off Callao, under the command of Sir Thomas Cochrane, afterwards tenth Earl of Dundonald. In January, 1820, he stormed the forts of Valdivia, and during the next two years swept Spain's naval forces from the Pacific. In 1821, Peru declared her independence, and the advent of a Colombian army under Bolivar, who came to Peru in September, 1823, was followed by the decisive battle of Ayacucho in Peru, which ended finally Spain's long period of dominion in Central and South America. Bolivar finally left Peru in 1826.

We have one hint as to Trevithick's doings during his absence from the mines. In 1867 Mr George Hicks, of Newquay, made a journey to Peru,[1] and in the mining district of Pasco he met Richard Spry (b. 1793), a lead smelter, who had gone out

[1] Letter to Editor, *Mining Journal*, March 8th, 1884.

there with others under a Richard Vivian. They arrived at Lima November 5th, 1819, just when Trevithick's troubles were at a climax. Spry "only saw Trevithick once or twice", because the latter "had sailed off on a brig along the coast to be out of the way", while Spry in compliance with his contract had to leave for the interior. Spry continued: "We left the sea with heavy hearts and sad forebodings". Sad to relate Spry never saw England again.

When Trevithick returned to the mines, with the cessation of opposition to him, and backed by Don Pedro Abadia, the most influential of the adventurers in the Company, who always seems to have treated him with consideration and confidence, he had every prospect of bringing the affairs of the Company to a prosperous issue. But here, as in the case of the Thames Archway, he was bludgeoned by circumstances over which he had no control, for now active hostilities penetrated even to the mountain district of the mines, and on December 11th, 1820, a battle was fought at Pasco itself in which the patriot forces were victorious. The result of these distractions was the utter ruin of his hopes. The patriot forces broke up his machinery and did all the damage to the mines that they could, because they were regarded as a source of supply to the Spanish forces.

Miers says:[1]

When in Lima, I was told by Abadia's Mother-in-law Yavaria, that at the end of seven years, the engines had succeeded, in 1821, in draining the mines to the desired depth; but they had hardly time to commence the mining operations, when the patriot forces advanced, took possession of the mining district, and seized whatever property could be found on the spot. All those parts of the steam engines which were likely to be destroyed, robbed, or carried off, were carefully concealed in some hiding-place where they have since remained.... The concern was thus circumstanced when I was in Lima at the end of 1823, and the subsequent convulsed state of the country has prevented all farther proceedings.

Miers made the acquaintance of Abadia, his wife and brother, and received much information as to these events from them.

[1] Miers, *loc. cit.* II, 441.

"Abadia and Arismendi had wholly lost the sum of 600,000 dollars."

The fragmentary nature of the information we have as to Trevithick's doings from 1819 till the day he left Peru does not permit of arrangement in chronological sequence. Not only was the mining enterprise ruined by the war, but as he tells us, his own property was seized and he himself temporarily forced into the insurgent army. We are in doubt, even, as to when he left the country, but as he was well acquainted with Bolivar, who does not appear to have arrived in Peru till September, 1823, it would seem that it was in 1824 that Trevithick left the distracted country upon whose shores he had so hopefully landed some six or seven years earlier, his golden prospects having vanished in ruin and destruction. We have one glimpse as to the course of events at the mines after Trevithick's departure:[1]

Yesterday I rec^d a letter from Peru from Cap^n Hodge who says that the engines sent there to the pasco mines in the room of mine that was destroy'd by the spaniards are nothing more than playthings and the sheet copper pumps that was brazed have all burst and the english agents all turned oute, drunkards and robbers, and the scheam have fallen to ground with a total loss of all the property with oute even doing any thing. So much for London engineering and mining knolidge! He says that in a few months he shall return to England.

This is corroborated by the account given by two travellers who visited Pasco in 1834, when mines were being worked in Colquijilca:[2]

The English Company that commenced working mines here in 1827 or 1828 completely failed; one of the steam-engines they erected is entirely destroyed, the other, though standing, is in such a dilapidated state, that to be again rendered serviceable, it would be necessary to renew one-half of the machinery, and repair the boilers. In the course of a short time, if left in its present state, it will be entirely ruined.

It would appear that more than one attempt was made to

[1] *Enys Papers.* Letter to Giddy, dated Highgate, Lauderdale House, 1830, Dec. 21st.

[2] *Narrative of a journey from Lima to Parà across the Andes and down the Amazon in* 1834. By Lieut. W. Smyth and Mr F. Lowe, late of H.M.S. "Samarang". London, 1836.

rework the mines, and that some of Trevithick's engines at any rate were reinstated, for we have information that two of them were at work in 1872, when they were described by William Williams, who then saw them:[1]

On the 3rd March, 1872, I saw in Yauricocha Mine two of Mr Trevithick's engines at work; one of them was a horizontal 12-inch open-top cylinder pumping engine, about a 4-feet stroke; there were two fly-wheels about 10 feet diameter and a cog-wheel 7 feet diameter, giving motion to two wrought-iron beams working a 10-inch pump bucket. The other was a 12-inch cylinder winding engine with a large fly-wheel. Three Cornish boilers, about 5 feet 6 inches diameter, with 3 feet 9 inch tube, 30 feet long, made of $\frac{7}{16}$ths of an inch plates, supplied steam of 40 lbs. on the inch.

We can only suppose that a new company had been formed and had obtained a concession from the republican Government to work these mines.

We have seen that during the year 1817 or later, Trevithick visited Chile and opened up copper mines there; they were still being worked in 1872, when his son published his *Life*:[2]

The late Mr Waters, an eminent Cornish miner, who for many years managed some of these mines in the neighbourhood of Valparaiso, said that Trevithick's name was better known to the miners there than to the miners in Cornwall. This statement was made in the Dolcoath account-house at a public meeting, the speaker and the writer being both on the committee of management.

Simon Whitbarn, of St Day, informed the writer that at Copiapo and at Coquimbo he had seen large heaps of copper ore, apparently unclaimed, which the people said had been raised by Don Ricardo Trevithick. About 1830 a miner, returned from South America, made a claim for wages for watching mineral left behind by Mr Trevithick.

Although it was the precious metals which had been usually sought in South America, copper had also been obtained, though only to a small extent, both by the natives before the conquest and by the Spaniards subsequently. Since Trevithick's day the output of copper from the mines of Chile has increased till that country holds second place among those producing that metal,

[1] *Life*, II, 220.   [2] *Life*, II, 250.

and it is reasonable to claim that he should be included among the pioneers who have brought about this result.

A good story is told by Thomas Edmonds, to whom it was related by Trevithick himself, of an event which occurred at some time during his occupation among the mines, where or when we have no intimation:[1]

Upon one occasion Captain Trevithick was called upon to act in a novel capacity, that of a surgeon. An accident happened to a native engaged in working an engine erected at a place distant about two hundred miles from Lima, by which accident both of his arms were crushed. There was no medical man within the distance of two hundred miles, and Captain Trevithick, believing that death would ensue if amputation was not immediately performed, offered his services, which were accepted by the patient. The operation, he informed me, was successful; the man rapidly recovered, and showed a pair of stumps which could have hardly been distinguished from the result of an operation by a regular surgeon. It is not improbable that in the warfare in which he had been engaged Captain Trevithick had been present and assisted at amputations of limbs of wounded soldiers. He thus probably acquired sufficient confidence to undertake and perform the operation himself.

We have now to follow our adventurer to other fields and will conclude this phase of his story by quoting a letter written to Francis Trevithick by James Liddell, a lieutenant in the British Navy, who met Trevithick at Callao in 1822:[2]

Forty-seven years are now passed since I had the great pleasure of meeting your father in Peru, and I have a vivid remembrance of the gratification afforded to my mess-mates when he came to dine with us on board H.M.S. "Aurora", then lying in Callao. I was then a lieutenant of that beautiful frigate, and was introduced to your father by Mr Hodge, of St Erth, with whom I had become acquainted in Chili. I remember your father delighting us all on board the "Aurora" by his striking description of the steam-engine, and his calculation of the "horse-power" of the mighty wings of the condor in his perpendicular ascent to the summit of the Andes. Your father's strong Cornish dialect seemed to give an additional charm to his very in-

[1] *Life*, II, 271.
[2] *Life*, II, 248. Letter dated Bodmin, 1869, Nov. 3rd.

teresting conversation, and my mess-mates were most anxious to see him on board again, but he left shortly after for the Sierra.

The Pasco-Peruvian mines were those which your father was engaged to superintend before he left England, and he had actually managed, by incredible labour, to transport one or two steam-engines from the coast to the mines, when the war of independence broke out, and the patriots threw most of the machinery down the shafts. This fearful war was a deathblow to your father's sanguine hopes of making a rapid fortune. About a year after this terrible disappointment (I think in 1822),[1] the "San Martin", an old Russian fir frigate, purchased by the Chilian Government, sank at her anchors in Chorillos Bay, ten miles south of Callao, and your father entered into an engagement with the Government in Lima to recover a large number of brass cannon, provided that all the prize tin and copper on board which might be got up should belong to him. This was a very successful speculation, and in a few weeks your father realized about 2500l. I remember visiting the spot with your father whilst the operations were carried on, and being astonished at the rude diving bell by which so much property was recovered from the wreck, and at the indomitable energy displayed by him. It was Mr Hodge, and not I, who then urged in the strongest manner that at least 2000l. should be immediately remitted to your mother. Instead of this, he embarked the money in some Utopian scheme for pearl fishing at Panama, and lost all!

I had the honour of dining with Lord Dundonald on board the crazy frigate "Esmeralda", which carried his flag in Callao Bay, but I never heard of the gallant conduct of your father in swimming off to his ship and advising him of an intended assassination. I fancy that this must have occurred before I came on the station, probably in 1820, or 1821.

Here we would pause to say a few words in reference to Mrs Trevithick. From her letter quoted above we can see what a loyal helpmate she was. This is borne out by such entries in Trevithick's accounts as "Jane paid R. Jeffrys while I was in Wales", and "Cash of Jane at Sundy. times", which show that she attended to his business in his absence. Among the accounts is one, evidently written in her hand, with Elizabeth Tyack, a sister of Trevithick who carried on a considerable business in

---

[1] Stevenson states that the "San Martin" sank on July 16th, 1821.

her own name. In this account we find a medley of engineering, mining and private items. It debits her with a balance brought forward of more than a thousand pounds, as well as with sums for interest, old copper, brown rolls, silk, lace, sheeting and old boiler plates; while it credits her with £551. 13s. 3d. due to her from Wheal Francis, and with goods including sheet lead, sal ammoniac, duck, etc., "for Mr Whitaker's, Captn Barrett's and Mr Roger's engines"; "by a rule per Jas. Sanders"; with bar iron and boiler plates, and "Sundries on our own acct." In contrast with this carefully made out account is one in Richard's own hand with this same sister, which ends almost pathetically, "Settled according to Sister Tyack's book". How easily we can picture the vain struggle to obtain a balance thus abandoned in desperation! We feel that Jane must have steered her erratic husband clear of many a shoal and avoided many a pitfall. Had she always kept his books we should have obtained much more definite and interesting matter than we can now extract from them. But although these present such a medley of facts and figures, they occasionally throw a little light on family affairs, and are even here and there illuminated by a touch of humour, as in an account with his brother-in-law Henry Vivian, where we find "To cash allowed by Tommy for the cow including the calf", and "By grass for the half of a cow"; surely two of the quaintest items to be found between the boards of a dusty ledger. "Tommy" was Thomasina, another sister, and Vivian's wife.

From the brief glimpses we get of their married life we see Jane Trevithick, a capable and intelligent woman, devoted to her husband and family, full of affection for him and admiration for his abilities and the better qualities that we know he possessed, while silent, possibly with occasional bitterness, as to the great defects that also appear in his character. Of him she said that[1]

her husband was good-tempered, and never gave trouble in home affairs, satisfied with the most simple bed and board, and always busy with practical designs and experiments from early morning until bed-time. He sometimes gossipped with his family on the immense

[1] *Life*, I, 239.

advantages to spring from his high-pressure engines, and the riches and honour that would be heaped on him and his children, but thought little or nothing of his wife's intimation that she hardly had the means of providing the daily necessaries of life.

And we see this all through Trevithick's career, he "never is, but always to be, blessed". Caring nothing for the past and little for the present, his eyes were ever fixed on the future that owed its golden hue to the wealth of his optimism.

Jane's affection and admiration for her husband come out strongly in her letter above cited, when she was roused like a lioness at bay, in defence of her absent and neglectful mate. Francis Trevithick states positively[1] that during the eleven years of absence in South America his wife and family received no assistance from him. We have seen by his letter to James Smith, cited above, that Trevithick asked the latter to insure his life. He, however, made no provision for the payment of the premium, to meet which his wife was compelled to sell her small property.

The only extenuation or explanation that can be offered for this neglect is that previous to his sailing to South America he had left the pole engine business in the hands of William Sims, his co-partner in the patent, and that he expected this to produce an income, for of all his schemes this was perhaps the one in which he had most faith. Some small amount was thus forthcoming, but very little. We have a hint among the draft letters that Mrs Trevithick had some authority to deal with matters under the patent, perhaps a power of attorney to act on her husband's behalf.

In the early days of his absence abroad, Trevithick appears to have kept his wife informed as to his doings. In his letter to Smith quoted above, he requests that a copy of Page's letter shall be sent to her. A copy of a letter to her from W. Teague, apparently written in 1818, says:[2] "...I was agreeably informed that you had just received a letter from Lima & should be extremely glad to hear the news...". Letters between those interested in the Peruvian venture appear to have been passed on

[1] *Life*, II, 278.   [2] *Draft Letter Book.*

to her for perusal. After the great debacle Trevithick seems to
have neglected to write either to his wife or his friends at home.
That she did her best to keep up correspondence with her
husband is shown by a letter to Giddy asking him to forward one
to Abadia. She says:[1]

> I feel the greatest anxiety...respecting my Husband.... The South
> Whalers will shortly sail and you would further oblige me by inclosing
> the letter (if it meet your approbation) to Mr Du Bois, No. 12 Copthall
> Court near the Bank, with a request that he will forward it by the
> first opportunity to Lima.

But perhaps the most striking testimony as to Trevithick's
character and at the same time to the depth of his own regard for
Trevithick himself is given in a letter from Giddy written in
1827. The county gentleman, the Member of Parliament, the
President of the Royal Society, writes to Trevithick admonishing
him in kindly terms such as indicate the confidential friendship
that had existed so long between them. He says:[2]

DEAR TREVITHICK,

Although many years have now elapsed since any direct communications have reached me from you, or since those who had much
stronger reasons for hoping that you would not neglect to inform them
at least of your proceedings, have known any thing about them: yet
I entertain a firm opinion of your still continuing the same honest,
thoughtless, careless man that I ever knew you, and that in the event
of such success attending your exertions as would prove satisfactory
to your own mind, that you would return and share Prosperity with
those most nearly connected and most entitled to your kindness and
protection. But while this uncertain attainment is in progress, Human
Life has advanced and is wearing away. Mrs Trevithick is advanced
beyond the middle Period of Life, your children are become Men and
Women—and their very support and maintenance has been owing to
the kindness of Mr Harvey.

I believe that no Woman ever conducted herself in a more exem-
plary manner than Mrs Trevithick during the whole of your absence
or with greater care and attention to her children. Suppose only the
case of her having abandoned them! And some years since when a

[1] *Enys Papers.* Letter dated Penzance, 1819, May 4th.
[2] *Enys Papers.* Press copy, dated 45, Bridge Street, Westminster, 1827,
Feb. 11th.

report was current in the West of Cornwall and generally believed,
that you had a second Family in South America, Mrs Trevithick
declared to me in the strongest Terms that she never did, and never
would believe it: but on the contrary she promised herself if you were
successful that she herself and her Family would partake of it with you.
I enclose her two letters, the second was written in consequence of my
having in answer to the first recommended the plan of sending out one
of your Sons without delay....

...With every wish for your Prosperity and Happiness, Believe
me,

Your sincere Friend and Humble Servant,

DAVIES GILBERT.

It is extremely doubtful, however, if Trevithick ever received
this letter and its enclosures. It was written in February, 1827.
In October of that year he was back in England. Giddy sent
his letter to Mr Malcolm MacGregor, the British Consul at
Panama, who in politely acknowledging it states that he "had
not the pleasure of being personally acquainted with Mr Trevithick" and that he had forwarded the letter by the schooner
"Britannia" sailing on that day to the coast of Guatemala
"where Mr Trevithick has his establishment".[1] The worthy
Consul no doubt had acted on erroneous information, as we have
no knowledge that Trevithick ever visited that country. At this
time he was probably forcing his adventurous way through the
forests of Costa Rica to the Caribbean Sea, or was nearing Cartagena on his long journey home, as we shall presently see.

Giddy's proposal that one of Trevithick's sons should be sent
to join his father in South America is of interest in view of
Francis's remark when speaking of Trevithick's intention of
shipping ore from Caxatamba to England that "the writer, who
was to have returned to Peru in the ship that had been engaged to
carry it, lost the chance of being a youthful traveller in foreign
lands".[2] This must have been before Trevithick left Peru, and
before Giddy made the suggestion, but Francis, so far from
having arrived at man's estate, was yet a schoolboy when his

[1] *Enys Papers.* Letter dated Panama, 1827, July 6th.
[2] *Life,* II, 255.

father returned to England. We are willing to accept Jane Trevithick's belief in her husband's fidelity, as against that of the scandal-mongers of Cornwall, for there appears to have been not the least ground for the accusation.

Simon Goodrich in his Diary, now in the Science Museum, South Kensington, throws some light on her life in the interval, in a lively account which he gives of a visit to Cornwall in 1825. Mrs Trevithick had been looked after by her brother Henry Harvey and installed with her eldest son in the inn outside Hayle Foundry, where customers and visitors to the works were in the habit of putting up. Goodrich says:[1]

Mr Woolf had gone on to Hale Works about 4 miles further to the Westward whither I was to follow him. I mounted my Rosinante who gaily galloped off, Mr Woolf's servant on another Horse shewing me the way. Mr Woolf had ordered dinner for us at the Hotel at Hale kept by Mrs Trevithic, the Wife of the Engineer, she being the sister of Mr Harvey of Hale Works, and I was amused to see written up over the Door Richard Trevithic Junr. It does not appear that Trevithic's family have lately heard from him, he does not correspond with them properly which if he be alive I am sorry to hear. I was excessively thirsty when I arrived from the great perspiration I had undergone in the mine and drank off as the best thing a large tumbler of Brandy and water, and then set down to dinner.

Mr Harvey came and took a Glass of wine with us after dinner and then shewed me round his works which were formerly Copper Smelting Works and are now converted into an Extensive Iron Foundry and Steam Engine Manufactory; most of the great work for the Cornish engines being now executed here. I was surprised to see the extent of this work and the number of Tools and means for doing work considering what is doing in other parts of the country, and notwithstanding this there is a Rival Foundry and Engine manufactory in the neighbourhood of this. Mr Woolf employs this work in making his Engines. Mr Harvey was very civil to me in consequence, he said, of my having been so to him on occasion of his visiting the works at Portsmouth Yard but I did not recollect it till he put me in mind of it.

Returned late in the Evening to Camborne Mr Woolf taking the lead in his Gig and I close after him on Rosinante whose pace was

[1] Goodrich Papers, Science Museum, South Kensington, Diary, No. 26, October, 1825.

much mended for fear of being left behind and who being well fed
began to be more spirited.

Arrived safely at Knapps Hotel.

To change the scene again to South America; we next find
Trevithick at Guayaquil, the chief seaport of Ecuador. He was
on his way to Bogota in Colombia, as we have seen, on a mission
for Bolivar. At Guayaquil he appears to have met for the first
time James M. Gerard, a Scotsman, who is stated to have come
of a good family, and to have been disappointed in the expectation of succeeding to an inheritance. Probably he was of the
family of Gerard of Rochsoles in Lanarkshire and was a younger
son. Gerard was leading the life of a trader on the Pacific coast,
as a letter of his indicates. The author of the *Life*, who gives this
letter in full, does not state to whom it was written. Seeing that
Gerard left Costa Rica in January, 1823, before Trevithick could
have left Peru, it would seem that it was written at Guayaquil,
most probably to Trevithick himself. Gerard had visited the
Costa Rican mines shortly after they had been reopened and his
letter says:[1]

In the month of June, 1822, I disembarked in the port of Punta de
Arenas, in the Gulf of Nicoya, the only one corresponding to that
province at present in use on the Pacific side. My object was to dispose
of a cargo of cotton which I had brought from Realejo, and to purchase
sugar in return. Circumstances, not necessary to mention, and the loss
of the small vessel with which I was trading on the coast, caused me to
remain in Costa Rica. Its name implies a very early conviction of its
natural opulence; it is certain that gold and silver abounded among the
Indians at the period of its conquest by the Spaniards. It was at one
time a favoured and flourishing agricultural colony, but from various
causes sank into neglect. Such was the apathy, both of the Government and of individuals, that the very existence of the precious metals
in the country had been almost entirely forgotten. In the end of 1821,
a poor man, Nicolas Castro by name, opened the first gold mine
known in Costa Rica since the conquest; and his success soon induced
others to try their fortunes; with fortunate results, in a few months a
mining district sprang into being.

A gentleman of the name of Alverado constructed at a very considerable expense what is called an Ingenio, consisting of various

[1] *Life*, II, 260.

edifices for depositing the ore, machinery driven by water for grinding it and afterwards blending it with quicksilver for amalgamation.

When I had landed in June, 1822, only five or six mines had been discovered, but in January, 1823, when I left the country, I cannot pretend to enumerate those in a state of progress and of promise.

After describing the processes as carried on at the mines, he concludes:

As the ground begins immediately to spring from the coast, and does so indeed very rapidly, a few miles takes us beyond the region of even these slight fevers, and as we continue ascending to the central tableland, a climate is encountered that may vie with any in the world for benignancy and beauty. We there meet with the fruits of the torrid zone, and near them the apple and the peach of Europe. The orange tree is in bearing the whole year. As in all situations within the tropics, it has a proper rainy season, but it is less inconvenient and disagreeable than might be expected, for it seldom rains two days in succession, and when it does, is invariably succeeded by an interval of fine weather; for the most part every day presents a few dry hours. The mines are situated on the ridges of the Cordillera, which without presenting snow-covered peaks, attain, nevertheless, considerable elevation. The clouds constantly attracted by those high summits, render the rainy season more severe in the mining district than in the plains. The greatest inconvenience was from the snakes, which in those solitary jungles, now first invaded by man, are very numerous and many of them venomous. Provisions are cheap and excellent. In short, there is but one fault I find with the country, and it is a great one, I mean the frequency of earthquakes.

Trevithick, whose sanguine temperament could never be long damped even under the most crushing misfortunes, determined to throw in his lot with Gerard, and, relinquishing his engagement with Bolivar, we find him interesting himself keenly in the mines on the mountain plateau of Costa Rica. All the information we have as to the occupation of the two men during the time they were in the country is contained in one or two letters and in a memorandum and a report written on their return to England. The mines are situated on the plateau between the two main ridges of the Cordillera at about a third of the distance from the Gulf of Nicoya to the Caribbean Sea. Trevithick and Gerard spent some four years in the country, familiarizing themselves with the mining that had already been done. They acquired mining rights apparently of considerable value, and made surveys and obtained much information to enable them to work their concessions to advantage. The Memorandum, in Gerard's writing, says:[1]

Though the plans and sections explain themselves, a few observations will not be misplaced. The deep adit for the Coralillo would be 600 yards, that for Quebrada-honda 400 yards, and besides serving as drains would form admirable roads for conveying the ores into the vale where the stamps must be erected.

The veins would be worked upward from the adits, and thus no expense would be incurred for ages to come in lifting either water, ore, or rubbish to the surface. Padre Arias Mine is an exception, requiring a powerful water-wheel, or an hydraulic pressure-engine, for which there is a fine fall of water of 135 feet. The mines in Quebrada-honda are those in which an interest has been procured. Captain Trevithick has an interest in the mine of Coralillo; the great watercourse is also his.

It will be seen by the plan that there are 75 fathoms fall to the point where his present mill is situated, and other 75 fathoms to the junction of the rivers of Quebrada-honda and Machuca. The whole length does not amount to two miles, within which it is estimated that sufficient power may be commanded to stamp 500,000 of quintals [roughly 25,000 tons] annually. To bring it up to that pitch, the waters of Machuca must be brought to join those of Quebrada-honda at Trevithick's mill, and then 40 tons of water per minute could be delivered in the dry season.

The report above referred to, made by Trevithick and Gerard in 1827, gives a good deal of further information:[2]

This map consists of several distinct parts. The middle part shows the mining district, the present dimensions of which are small, the length being hardly four miles, breadth from two to three, and the superficial extent from eight to ten miles. The upper part of the plan is a section of the north ridge, called Quebrada-honda, and shows the line of the proposed adits. The lower part in like manner exhibits the south ridge, called Coralillo. The map further shows the inclination or gradual fall of the ground along the valley, and of the streams by which the mills are driven.

[1] *Life*, II, 262.   [2] *Life*, II, 263.

The canal is likewise shown 5000 yards in length, by which the rivers of Machuca would be brought to join that of Quebrada-honda.

Castro's mine is situated on the southern ridge, and was the first mine worked to any extent. There the veins are very large; in fact, from the manner in which a number of horizontal veins are seen falling into the perpendicular or master vein, the great body of the mountain would appear to consist of lodes. This mass of ore is in general rich. It has been worked open to the surface, somewhat like a quarry, so that it is not difficult to calculate in cubic feet the quantity that has been excavated. The mine is supposed to have yielded in the course of the last six years gold to the value of 40,000*l.*, and by measuring the excavations it would appear that this amounts to, on an average, one ounce of fine gold to every ten or twelve quintals of ore. In 1821 the existence of silver was only imagined. In 1823 it was fully ascertained. Ever since 1824 it has constituted a small but constant portion of the produce of Quebrada-honda, and in 1827 it was decidedly evinced in Coralillo. The discovery of gold in Coralillo led them to work in Quebrada-honda, where they found both gold and silver, and the discovery of silver in Quebrada-honda, by strengthening the expectation of it in Coralillo, led in its turn to the discovery of silver there. In Quebrada-honda they only work on the ground in the immediate vicinity of the stream, and that in the most imperfect manner; but great light has been thrown on the value of the ores on this spot and in the district generally by the progress made in working what is called Padre Arias Mine, which takes its name from an ecclesiastic who first worked it. This mine is situated in low ground near the verge of the stream, and was at first only worked for gold. There were soon, however, indications of silver, which increased progressively in sinking, till at the depth of only 10 yards the influx of water exceeded the means of draining, and the works under water-level were necessarily abandoned, at a time when ores were yielding upwards of 200 oz. of silver to the ton, a striking proof of the tendency of silver ore to improve in this district as the depth increases.

Mr Richard Trevithick, that eminent Cornish miner and engineer, so well known for his inventions, and particularly for the high-pressure steam-engine and the drainage of the Pasco Mines in Peru...having heard favourable reports of the mining district we are now describing, soon after repaired thither, and was so fully impressed with its value and importance that he made an extensive contract for different properties, and resided in the country for four years.

He is now in England ready to give explicit answers to any inquiries that may be made as to the mineral wealth of Costa Rica, and the extraordinary facilities afforded by its position and natural advantages. An estimate has been made for establishing a complete mining concern in Costa Rica, with houses, iron railroads, stamping mills, &c., so as to raise, stamp, and bring into refined gold the produce contained in 250,000 tons of ore per year.

The report goes on to enumerate the advantages of the district, and concludes thus:

Captain Trevithick and Mr Gerard, with a particular view to the enterprise now under consideration, and after considerable risk and labour, succeeded in laying down the navigable head of the Serapique and in throwing such light on the intervening tract as will be of great assistance to future adventurers. They ultimately constructed a canoe in which they sailed down to the port of San Juan.

But we are anticipating the journey mentioned in the last sentence. Trevithick and his companion, being anxious to return to England to interest capitalists in their mining ventures, and wishing to avoid the long voyage around Cape Horn, the only route then available for the shipment of machinery to the mines and of their products to England, decided boldly to solve the problem themselves, by exploring a long-talked-of route overland to the Caribbean Sea; and accordingly started on a journey which, although its object was peaceful, recalls to our minds old stories of adventure and privation upon the Spanish Main. A pretty story might be written, but it will be most convincingly told in the brief account of the adventurers themselves.

A party of eleven started to make their way through the rugged passes and ravines of the Cordillera, and through tropical forests and swamps, to the eastern coast of Central America. Their company included two boys going to England to be educated. How tame must the routine of a school at Highgate have appeared to these lads, the commencement of whose long journey to a foreign land was conceived in the spirit of the Conquistadores themselves!

Thomas Edmonds in after years gave his recollections of Trevithick's own account of their adventures:[1]

In 1830 I frequently saw Trevithick at the house of Mr Gittins, at

---

[1] *Life*, II, 269.

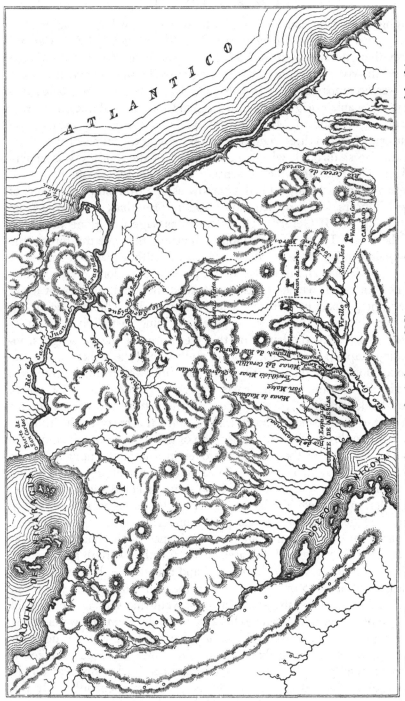

Fig. 31. Route followed by Trevithick across the isthmus of Nicaragua, c. 1826. From a map prepared by his son

Highgate, a schoolmaster, with whom were two boys that had accompanied him from Costa Rica, called Montelegre. Before Captain
Trevithick no European had adventured on or explored the passage
along the river from the Lake Nicaragua to the sea. In the adventure
he was accompanied by Mr J. M. Gerard, a native of Scotland; two
boys of Spanish origin going to England for their education; a half-
caste, as servant to Mr Gerard; and by six working men of the
country, of whom three went back, after helping to remove obstructions
in the forest through which the first part of the journey was undertaken. The risks to which the party were exposed on their passage
were very great; they all had a narrow escape from starvation, one of
the labourers was drowned, and Captain Trevithick was saved from
drowning by Mr Gerard. The intended passage was along the banks
of the river. To avoid the labour of cutting through the forest, the
party determined to construct a raft, on which they placed themselves,
their provisions, and utensils; after a passage of no long duration they
came to a rapid, which almost overturned their raft, and swept away
the principal part of their provisions and utensils. The raft, being unmanageable, was then stopped by a tree lying in the river, with its
roots attached to the bank; on this tree three of the passengers, including Captain Trevithick, landed, and reached the bank; this was no
sooner done than the current drove the raft away from the tree, and
carried it, with the remaining passengers, to the opposite bank, where
they landed in safety, and abandoned the raft as too dangerous for
further use. The next object was to unite the party again into one
body. The three left on the other side of the river were called upon to
swim over: one of the men swam over in safety, the next made the
attempt and was drowned, the third and last remaining was Captain
Trevithick, who was either unable to swim or could swim very little.
In order to improve his chances of safety, he gathered several sticks,
which he tied in a bundle and placed under his arms; with these he
plunged into the stream; but the contrivance of the bundle of sticks
afforded him very doubtful assistance, for the current appeared to
seize the sticks and whirl him round and round. He, however, finally
reached within two or three yards of the bank in a state of extreme
exhaustion. Mr Gerard going into the water himself and holding the
branch of a tree, then threw to his assistance the stem of a water-plant,
holding one of the extremities in his own hand. It was not until the
fourth time of throwing that Captain Trevithick was able to seize the
very extremity of the plant (which was leaf) in his fingers; on the
strength of the leaf his life on the occasion was dependent. It was
determined to give up any further idea of using a raft on the river, and

to continue their journey along the banks of the river. For subsistence for the remainder of their journey they had to depend on the produce of one fowling-piece and a small quantity of gunpowder; after a few days the gunpowder got wet by accident, and in the attempt to dry it, it was lost by explosion. The party finally arrived in a state of great exhaustion at the village, now the considerable port of San Juan de Nicaragua, or Greytown.

Trevithick described the journey as "lasting three weeks, through woods, swamps, and over rapids; their food monkeys and wild fruit; their clothes at the end of the journey shreds and scraps, the larger portion having been torn off in the underwood".[1]

Having constructed rafts and canoes and followed the river Serapique to its junction with the San Juan, they ultimately reached the port of the same name at the mouth of that river.

Having now followed our erratic adventurers to the shores of the Caribbean Sea, we find ourselves at fault. All clue is lost, but we pick up the scent again some five or six hundred miles away from San Juan on the coast of the main continent of South America. The scene opens in an inn at Cartagena in Colombia, and is described by the chief actor in the incident: "In the public room at the inn, he was much struck by the appearance and manner of two tall persons talking English; the taller of them, wearing a large-brimmed straw or whitish hat, paced restlessly from end to end of the room".[2] How this incident brings to our recollection the scene at Soho, when Trevithick reluctantly faced the meat-pie! The narrator was young Robert Stephenson, then returning home from an engagement with the Colombian Mining Association, to become a railway engineer of world-wide reputation.

How Trevithick and Gerard had got to Cartagena we do not know, but we conclude that they had gone there to try to find passage to England. What we do learn is that near Cartagena Trevithick had had another hair-breadth escape, as the following

---

[1] *Life*, ii, 269.

[2] *Life*, ii, 274, but cf. J. C. Jeaffreson, *Life of Robert Stephenson*, i, 105, where a somewhat different account is given.

letter written to Mr Edward—afterwards Sir Edward Watkin—shows:[1]

I read in the public prints that in a speech made by you in Belle Vue Gardens you referred to the meeting of Robert Stephenson with Trevithick at Cartagena, which, if your speech be correctly reported, you attribute to accident. The meeting was not an accident, although an accident led to it, and that accident nearly cost Mr Trevithick his life; and he was taken to Carthagena by the gentleman that saved him, that he might be restored. When Mr Stephenson saw him he was so recovering, and if he looked, as you say, in a sombre and silent mood, it was not surprising, after being, as he said, "half drowned and half hanged, and the rest devoured by alligators", which was too near the fact to be pleasant. Mr Trevithick had been upset at the mouth of the river Magdalena by a black man he had in some way offended, and who capsized the boat in revenge. An officer in the Venezuelan and the Peruvian services was fortunately nigh the banks of the river, shooting wild pigs. He heard Mr Trevithick's cries for help, and seeing a large alligator approaching him, shot him in the eye, and then, as he had no boat, lassoed Mr Trevithick, and by his lasso drew him ashore much exhausted and all but dead. After doing all he could to restore him, he took him on to Carthagena, and thus it was he fell in with Mr Stephenson, who, like most Englishmen, was reserved, and took no notice of Mr Trevithick, until the officer said to him, meeting Mr Stephenson at the door, "I suppose the old proverb of 'two of a trade cannot agree' is true, by the way you keep aloof from your brother chip. It is not thus your father would have treated that worthy man, and it is not creditable to your father's son that he and you should be here day after day like two strange cats in a garret; it would not sound well at home". "Who is it?" said Mr Stephenson. "The inventor of the locomotive, your father's friend and fellow-worker; his name is Trevithick, you may have heard it", the officer said; and then Mr Stephenson went up to Trevithick. That Mr Trevithick felt the previous neglect was clear. He had sat with Robert on his knee many a night while talking to his father, and it was through him Robert was made an engineer. My informant states that there was not that cordiality between them he would have wished to see at Carthagena.

The officer that rescued Mr Trevithick is now living. I am sure he will confirm what I say, if needful.

[1] *Life*, ii, 272. Letter from James Fairbairn, dated Stanwix, Cumberland, 1864, Nov. 27th.

The officer was Bruce Napier Hall, to whom Sir Edward Watkin wrote, as suggested, and received the following reply:[1]

I was not aware that he [i.e. Mr Fairbairn] had written to you. He brought me a paper with your remarks about the meeting of Mr Robert Stephenson and Mr Trevithick, and asked me if it were true that they met at Carthagena as stated, as he [Mr Fairbairn] thought it was at Angostura, and that Mr Trevithick was in danger of being drowned at the Bocasses, i.e. the mouths of the Orinoco, the Apure, &c., &c., I explained that it was near the mouth of the Magdalena.

I will just say that it was quite possible Mr R. Stephenson had forgotten Mr Trevithick, but they must have seen each other many times. This was shown by Mr Trevithick's exclamation, "Is that Bobby?" and after a pause he added, "I've nursed him many a time".

I know not the cause, but they were not so cordial as I could have wished. It might have been their difference of opinion about the construction of the proposed engine, or it might have been from another cause, which I should not like to refer to at present; indeed, there is not time.

These two men wrote, it will be observed, thirty-seven years after the event; their description of the meeting puts Robert Stephenson in a somewhat invidious light, although in the authors' opinion, quite unjustifiably. At this time he was about twenty-four years of age. His father's acquaintance with Trevithick must have occurred at about the time that the Newcastle locomotive was constructed—say between 1804 and 1806. Robert was but an infant when Captain Dick dandled him on his knee, and a boy of thirteen years when Trevithick left England. As a matter of fact Robert Stephenson behaved most handsomely on this occasion: "he had a hundred pounds in his pocket, of which he gave fifty to Trevithick, to enable him to reach England".[2]

Stephenson and his colleague Empson (who was likewise returning from the Colombian mines) took passage in the brig "Bunker's Hill", bound for New York. Gerard parted from Trevithick, and with the two Montelegre boys sailed with

[1] *Life*, II, 273. Letter dated 4, Earl Street, Carlisle, 1864, Dec. 16th.
[2] *Life*, II, 274.

Stephenson. Their adventures were not at an end, however, for the brig was wrecked within sight of land, but fortunately every one escaped with his life. Gerard, whose movements were hampered by a scarcity of "the needful", was invited by Stephenson and Empson to accompany them on a visit to Niagara Falls before he returned to England, and he was glad to do so.[1]

And what of Captain Dick? He probably found his way from Cartagena to Jamaica, and thence sailed by a packet to Falmouth, but we have no record. The money given to him by Stephenson ought to have served to pay the passage, but his biographer says: "His passage money being unpaid, a chance friend enabled him to leave the ship". Whatever the actual facts, he landed penniless on the shores of his native Cornwall, his sole possessions being "the clothes he stood in, a gold watch, a drawing compass, a magnetic compass, and a pair of silver spurs"—if we except a hazy right in his Caxatambo copper-mountain and an option on the Costa Rican mine of Quebrada-honda. Thus was rung down the curtain on his great adventure.

[1] *Life*, ii, 279.

CHAPTER VI

LAST FLASHES OF HIS GENIUS

Changes during his absence in South America—Company promoting, Costa Rican mines—Gun carriage—Visits Holland on drainage schemes—Petition to Parliament for grant—Duty of steam engines—Patents for boilers and superheaters—Application to Admiralty for trial vessel—Domestic heating—Reform Act column—Work at Dartford—Illness, death, and burial—Character.

DURING the decade that Trevithick had spent in South America much water had flowed under the bridges. The first quarter of the nineteenth century was a period of rush in technical development in all directions, so hurried indeed that there was little opportunity to record it, and consequently we know perhaps less of it than of any period before or since. It was upon this changed and changing world, arising in some measure from the inventions he had himself introduced, that Trevithick now looked forth.

The high-pressure engine was in everyday use; indeed other prophets of high-pressure steam such as Jacob Perkins had arisen. The Cornish engine, supplied by Trevithick's boiler, now coming to be known as the Cornish boiler, performed a "duty" between three and four times that of the Watt engine, largely owing it is true to the friendly rivalry engendered by the systematic reporting of the duty; a type of engine suitable for paddle propulsion of vessels had been arrived at and deep-sea navigation had been attempted; the rail locomotive had been developed to the stage of hauling mineral traffic at slow speeds on a public steam railway; the construction of the Liverpool and Manchester Railway had been begun and passenger service by the locomotive was in sight; the steam coach was already plying on common roads at a speed equal, if not superior, to that of the horsed coach, only to be thwarted later, however, by short-sighted legislation; iron manufacture had progressed by leaps and bounds and the hot blast was being experimented with at

Clyde Ironworks. Old associates and friends had died, retired or occupied other spheres. For instance, his old friend Rastrick had taken up railway engineering, and Homfray had sought a wider field in politics. Tredgold, Farey and other writers had published works on the steam engine; Mechanics' Institutes had been founded by Dr G. Birkbeck in 1824, education had become more general and a new generation had come forward. And so we might go on enumerating changes in these and in other fields than those in which Trevithick had worked, but enough has been said to show that he had much leeway to make up.

Cornwall alone seemed the same. Trevithick was welcomed home at Hayle as if he were a conquering hero, and not a penniless adventurer,[1]

by all the neighbourhood by ringing of bells, and entertained at the tables of the county and borough members, and all the first-class of gentlemen in the west of Cornwall, with a provision about to be made for me for the past services that this county has received from my inventions just before I left for Peru, which they acknowledge to be a saving in the mines since I left of above 500,000*l*., and that the present existence of the deep mines is owing to my inventions. I confess that this reception is gratifying, and have no doubt but that you will also feel a pleasure in it.

His son Francis gives a graphic account of his father's arrival:[2]

In the early part of October, 1827, the writer, then a boy at Bodmin school, was asked by the master if any particular news had come from home. Scarcely had the curiosity of the boys subsided, when a tall man with a broad-brimmed Leghorn hat on his head entered at the door, and after a quick glance at his whereabouts, marched towards the master's desk at the other end of the room. When about half-way, and opposite the writer's class, he stopped, took his hat off, and asked if his son Francis was there. Mr Boar, who had watched his approach, rose at the removal of the hat, and replied in the affirmative. For a moment a breathless silence reigned in the school, while all eyes were turned on the gaunt sun-burnt visitor; and the blood, without a defined

---

[1] *Life*, II, 277. Letter to J. M. Gerard, dated Hayle Foundry, 1827, Nov. 15th.

[2] *Life*, II, 276.

reason, caused the writer's heart to beat as though the unknown was his father, who eleven years before had carried him on his shoulder to the pier-head steps, and the boat going to the South Sea whaler.

We are able to give the exact date of Trevithick's arrival, for it is recorded in a letter from Davies Giddy to his daughter Kitty; the tone of it distinctly suggests the return of the prodigal:[1] "I must add that on last Tuesday [i.e. Oct. 9th] Capⁿ Richard Trevithick, after an absence of eleven years, during the whole of which he has not held any communication whatever with his Family, arrived suddenly at Hayle".

Doubtless it was sweet to Trevithick, after his Ulysses-like wanderings, to breathe once more the air of his native country, but the rest could not last long, for it was necessary to look round for ways and means even had he not, fired by what he now knew of the mineral riches of Central America, determined, in association with his friend Gerard, to exploit the mines in Costa Rica and Chile. To put him in command of funds he was sanguine that he would obtain tangible recognition of the great savings that had accrued to the mining interests of Cornwall through him, by the adoption of the Cornish engine and boiler. Had he reflected a moment, he would have recalled how difficult it had been for Boulton and Watt to collect their royalties, and how he himself had helped to create those difficulties, even when, as was not Trevithick's case, they had legal power to recover.

However, Trevithick still had claims that were morally if not legally enforceable. The day before setting out for South America he had agreed to sell half of his plunger-pole patent to William Sims (who was really only agent for Michael Williams of the United Mines—subsequently M.P. for West Cornwall) for the sum of £200 to cover the counties of Devon and Cornwall.[2] Richard Edmonds, Trevithick's solicitor who drew up the agreement, made the following note among the papers relating to the transaction:

Mr Michael Williams said it was verbally agreed that Capt.

---

[1] *Enys Papers.* Letter dated Tredrea, 1827, Oct. 14th.
[2] *Life*, II, 106, 107. Letter from Richard Edmonds, dated 1853, Feb. 8th.

Trevithick should have one quarter part of the saving above twenty-six millions.

Edmonds goes on to say:

This, I believe, was the average duty of the engines at that time. I had several assurances relative to Trevithick's claims, and much correspondence, but no allowance was made from any mines but Treskerby and Wheal Chance; though Trevithick's patent and boilers were used throughout the county without acknowledgment; and the duty of the engines had soon increased from twenty-six millions to about seventy millions.

In 1819 I attended at the account-houses of Treskerby and Wheal Chance, of which the late Mr John Williams, of Scorrier, was the manager, in consequence of some of the adventurers objecting to continue the allowances on the savings to Captain Trevithick, when Mr Williams warmly observed, that whatever other mines might do, he would insist, as long as he was manager for Treskerby and Wheal Chance, the agreement made should be carried into effect.

Edmonds further stated in 1853:[1]

The agreement was that patentees should have one-fourth part of the savings of coal above twenty-six millions. The one-half of this fourth part from these two mines for some years was about 150l. per annum. This did not relate to the boilers; Trevithick unfortunately did not take out a patent for that improvement. The adventurers of two or three mines only had the honesty to pay 100l. for each mine; others made use of it without acknowledgment.

That Mrs Trevithick did actually receive some money is confirmed by her correspondence with Giddy, whose help she sought in obtaining payment for the United Mines engine:[2]

I beg the favour of your writing to Mr John Williams Junr respecting the savings of the Engine on Mr Trevithick's plan at the United Mines. When my Brother meets him, he promises a speedy settlement, and when he has written to him he has given him an evasive answer. This has been the case three years; to his last letter, however, he has received no reply. I also have written to him, but with no better effect. Mr Mich^l Williams regularly pays the savings on the Treskerby Engine and I had hoped the Agents of the other Mines would have been induced to follow his example. Mr Sims is the acting

[1] *Life*, II, 305. Letter dated Penzance, 1853, Jan. 12th.
[2] *Enys Papers*. Letter dated Phillack, 1820, Oct. 7th.

partner and he being entirely in the power of Messrs Williams prevents one having recourse to Law.

He wrote on her behalf to Mr John Williams, who replied as follows:[1]

With regard to Mrs Trevithick's claims for savings on engines at the United Mines, there is much to be said.

Before Mr Trevithick went abroad he sold half the Patent right to William Sims, our Engineer, who very strongly recommended that two of the engines at the United Mines should be altered to what he considered his patent principle, but the alterations proved very inferior to his expectations, and to this circumstance I attribute much of the objections in question. Mr Henry Harvey has perhaps told you who the partners are in the Patent, and when you next come into this County I shall be much pleased to wait on you at Tredrea that you may hear the whole of the case; and tho' the United Mines Advrs are far from being a United Body, I am sure my Sons, who are their Managers, are desirous to recommend what appears to them right, and they will with myself be obliged for your Opinion after you have heard the whole matter on both sides.

We can be sure that Giddy's opinion would be in Trevithick's favour on the grounds of justice, but Trevithick's son states positively that no payment was made. When Trevithick got to know about it after his return he seems to have been very angry and took an early opportunity of calling on Mr Williams. Francis Trevithick says[2] that his father "called at Scorrier House in a very threatening attitude on 31st October, 1827, when Mr Williams, sen., said the reason for not continuing the payment was his belief that the term of the patent had expired".

A lawyer's letter from Edmonds followed on November 7th, pointing out that the patent did not expire till May, 1830. The matter was closed by Trevithick's acceptance of the sum of £150 as a final sum, as reported in the following letter to Edmonds:[3]

Yesterday I called on Mr Williams, and after a long dispute brought the old man to agree to pay me 150*l.* on giving him an indemnification in full from all demands on Treskerby and Wheal Chance Mines

---

[1] *Enys Papers. Life,* II, 304. Letter dated Scorrier House, 1820, Nov. 14th.

[2] *Life,* II, 305.

[3] *Life,* II, 305. Letter dated 1828, Jan. 24th.

in future. He requested that you should make out this indemnification. I could not possibly get them to pay more, and thought it most prudent to accept their offer rather than risk a lawsuit with them.

Of any outcome of Trevithick's sanguine hopes of receiving half a million of money from the Cornish mines—or from any-where else for that matter—for his improvements we have found no trace, for he never received a penny piece. Nothing rings truer than Mr Williams's observation in a letter written in 1853:[1]

Mr Trevithick's subsequent absence from the county, and perhaps a certain degree of laxity on his own part in the legal establishment and prosecution of his claims, deprived him of much of the pecuniary advantage to which his labours and inventions justly entitled him; and I have often expressed my opinion that he was at the same time the greatest and the worst-used man in the county.

The arrival at Liverpool about November 11th of his friend Gerard, whom he had left at Cartagena with the two sons of Montelegre, and whom he had given up hope of seeing again, fired Trevithick to take up the scheme they had meditated together of mining in Costa Rica and for which they had both received concessions from the Government of that country. Trevithick at once invited him to come down to Cornwall to talk over their programme:[2]

I should be extremely happy to see you down here; it is but thirty-six hours' ride, and it will prepare you for meeting your London friends, as I would take you through our mines and introduce you to the first mining characters, which will give you new ideas and enable you to make out a prospectus that will show the great advantages in Costa Rica mines over every other in South America. I think it would not be amiss for you to bring with you a few specimens, and after you have seen the Cornish mines and miners I doubt not but we shall be able to state facts in so clear a light that the first blow well aimed will be more than half the battle, and prove a complete knock-down blow, which in my opinion ought to be completed previous to your opening your mining speculation in general in London.

[1] *Life*, II, 109.
[2] *Life*, II, 277. Letter already quoted, dated 1827, Nov. 15th.

Gerard, as soon as he reached London, replied as follows:[1]

I arrived here from Liverpool last night, and this morning had the pleasure of receiving your kind letter of the 15th. The brig "Bunker's Hill", in which we came from Carthagena to New York, was wrecked within a few hours' sail of the port. We were in rather a disagreeable situation for some time, but more afraid than hurt. The cargo was nearly all lost. The ship was got off, but a complete wreck. The cause, however, of my delay in arriving arose from the want of the needful. You recollect Mr Stephenson and Mr Empson, agents for the Colombian Mining Association, whom we met at Carthagena. They kindly offered to supply me, but having determined to visit the celebrated Falls of Niagara, they insisted on my accompanying them, which I did.

I am truly rejoiced to learn that your countrymen retain so lively a sense of the importance of your services. I think with you that before sounding the public or proceeding further, it might be well we should meet quietly to talk over everything and arrange our ideas, and that Cornwall, for the reasons you mention and others, would be the better place.

The boys are well, and desire their respects to you.

Gerard only delayed long enough in London to find a school for the two Montelegre boys. This he succeeded in finding at Highgate, then a pleasant village on the northern heights overlooking London. The school was the well-known "gents' boarding academy" kept by Mr William Gittins at Lauderdale House, a fine Queen Anne mansion fortunately still standing. Subsequently, when in London, Trevithick would seem to have resided there, for much of his correspondence bears that address.

Gerard then set off for Cornwall, and on arrival there Trevithick and he appear to have spent some busy weeks in drawing up reports and a prospectus of the Costa Rican mines with the obvious intention of coming before the public as company promoters. On December 22nd he and Gerard apparently left Hayle for London, and if so Trevithick must have spent his first Christmas in England away from home. Trevithick was soon back again and used his influence to get adventurers in Cornwall to take shares.

[1] *Life*, ii, 279. Letter dated 1827, Nov. 17th.

Gerard wrote to Trevithick:[1]

I had very unexpectedly a letter from Costa Rica this morning by the way of Jamaica, including two for you, which I have the pleasure of transmitting. Mine is from Montelegre, begun on the 25th of August, and finished on the 11th of September, when Don Antonio Pinto, with some people from the Alajuela, was to start by the road of Sarapique on his way to Jamaica. His intention was to find a better route as far as Buona Vista, after which he would probably nearly follow our course to the Embarcadero of Gamboa.

Whether he succeeded in finding a less rugged road to Buona Vista I do not know. That he reached his destination seems clear from our letters having come to hand; but from their old date it would appear that he had either met with difficulties on the road or with considerable detention at San Juan. Montelegre writes me that Don Yonge had effected a compromise on your account with the Castros. Gamboa got back to San José on the 18th August, twelve days after he parted from us, to the great joy of our mutual friends. Mr Paynter had been unwell after our departure. Both he and Montelegre desire their kindest recollections to you.

Thus all seemed to be going well at the Costa Rican end. Trevithick replied:[2]

Yesterday I saw Mr M. Williams, who informed me that he should leave Cornwall for London on next Thursday week, and requested that I would accompany him. If you think it absolutely necessary that I should be in town at the same time, I would attend to everything that would promote the mining interest. When I met the Messrs Williams on the mining concerns some time since, they mentioned the same as you now mention of sending some one out with me to inspect the mines, and that they would pay me my expenses and also satisfy me for my trouble with any sum that I would mention, because such proceedings would be satisfactory to all who might be connected in this concern. I objected to this proposal on the ground that a great deal of time would be lost and that the circumstances of your contracts in San José would not admit of such a detention; for that reason alone was my objection grounded, and if that objection could have been removed I should have been very glad to have the mines inspected by any able person chosen for that purpose, because it would not only take off the responsibility from us, but also strengthen our reports, as

[1] *Life*, II, 280. Letter dated 42, St Mary Axe, 1828, Jan. 13th.
[2] *Life*, II, 281. Letter dated Hayle Foundry, 1828, Jan. 30th.

the mining prospects there will bear it out, and that far beyond our report. Some time since I informed you that I had drawn on the company for 100*l.* to pay 70*l.* passage-money, and would have left 30*l.* to defray my expenses returning to London. The time for payment is up, but I have not as yet heard anything about it, therefore I expect there must be an omission by the bankers whose hands it was to have passed through for tendering it for payment. Perhaps in a day or two I shall hear something about it.

This suggests that Trevithick and Gerard had committed themselves to return at once to Costa Rica with mining plant and miners to implement their concessions, and that there was a time limit that would have expired before independent engineers could have been sent out to inspect and report; on the other hand, it was hardly to be expected that adventurers would "buy a pig in a poke". An incident typical of Trevithick's downright character occurred during the negotiations:[1]

At a meeting of several gentlemen in London, a cheque for 8000*l.* was offered to Trevithick for his mining grant of the copper mountain in South America. Words waxed warm, and the proffered money was refused. The next day Mr Williams said to him, "Why did you not pocket the cheque before you quarrelled with them?" Trevithick replied, "I would rather kick them down stairs!"

Alas! all negotiations failed, and Gerard, who had gone to France and Holland to find support in floating the projected company there but had met with no better success, died in Paris, in poverty. Thus vanished Trevithick's last remaining hope of riches coming from the New World.

But Trevithick since his return had been busy in another direction, viz. on a new method of loading broadside guns for men-of-war. Doubtless he had been impressed during the chilean war by the cumbrous and slow manipulation necessary in such muzzle-loaders. Briefly, his gun was mounted on trunnions running up slides at an angle of 25° to absorb the recoil, as well as to prime and cock the gun, while for loading the gun had to be turned into the vertical position by means of a

[1] *Life,* ii, 283.

lever; it returned to the firing position by gravity. Trevithick intended to have applied for a patent for the mounting, because he made a declaration on November 10th, 1827, and a specifica-

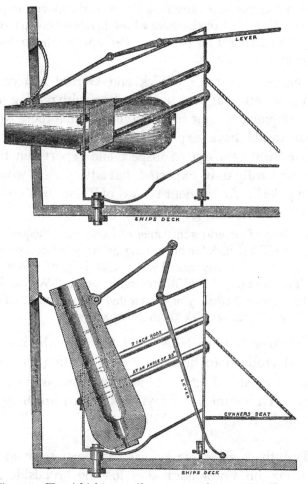

Fig. 32. Trevithick's recoil gun mounting, 1827. From a drawing by his son

tion[1] was drawn up; no further steps were taken to apply for a patent; probably he was deterred from proceeding to the Great Seal because of the unfavourable reception of his invention.

Trevithick wrote on February 21st, 1828, to Thomas Coch-

---

[1] *Life*, II, 284.

rane, Earl of Dundonald, who had just arrived back in England after his famous Bolivian exploits, to try and interest him in the scheme, but without success.

Trevithick entrusted a model of the gun and a drawing of an iron ship to be propelled by steam to his friend Gerard for submission to the Ordnance Board, or as it was then known, the Select Committee of Artillery Officers. They "reported[1] that on examination of the invention they consider it to be wholly inapplicable to practical purposes". Thus another hope was dashed. The report is what one would expect from practical gunners about something novel, and no doubt they were right, but we ought not to overlook the fact that Trevithick's was one of the first attempts to get useful work out of the recoil, as has since been so widely practised.

In the letter to the Earl of Dundonald, mentioned above, Trevithick says "that an iron ship was being laid down on a slip at Hayle Foundry" and there is evidence that it was actually commenced, but we hear nothing further about it. That he was able to get work done there must have been due to the fact that the proprietor, Henry Harvey, was his brother-in-law.

In June, 1828, Trevithick was in London, apparently in the rôle of consulting engineer. The following letter will show what he was doing:[2]

A few days sence a Mr Linthorn call'd on me, and requested me to accompany him to Cable Street, near the Brunswick theatre, to see a crane worked by the atmosphire, on a double acting engine attach'd to this crane. He have taken oute a patent[3] for the same, and has entred into a contract with the Saint Cathrine Dock Compy to work their cranes, 140 in number round their Dock, by means of a steam-engine of sufficient power to comand the whole, by air pipes being laid all round the docks, and a branch pipe from the main pipes connected to each separate crane. To each crane is fixed a ten inch cylinder,

---

[1] *Life*, II, 287. Letter from Master-General of the Ordnance, dated 1828, Feb. 21st.

[2] *Enys Papers. Life*, II, 292. Letter to Giddy, dated 1828, June 18th, addressed from 42, St Mary Axe, Gerard's office.

[3] Linthorn took out no patent; the patentee was John Hague, No. 5546, Aug. 30th, 1827.

double acting of 20 inches stroke, and the atmosphire acting on each side of the piston like steam in a double engine, is taken back through these Iron pipes to a large steam engine that works a large air-pump to extract the whole of the air from the whole of the 140 Double air engines attached to every separate crane. On being requested to give my opponion on this plan, after seeing one crane work'd, I inform'd them of the dissappointment that the Iron master, Mr Wilkinson, in Shropshire, severel years since expeainced, on the resistance of air passing through pipes of a long distance from his blast engine to his furnace. He told me he was aware of that circumstance, and it had been farther proved sence in London by one of the gass compainys who had attempted to force gass to a very considerable disstance, and who had also failed on the same grounds. But that forceing an elastice fluid, and drawing it by a vaccum, was a verry diffrent thing, and that the error found by forceing in steed of drawing was compleatley removed. For my own part I am not convinced on this head, but am of opponion still that the result on tryal will be found nearley the same. However, let that be as it may, the expence and complication of the machine having a double engine, with its gear attached to every separate crane, and the immence quantity of air that must be thrown in to the large air-pump from 140 Double engines of 10 Inches Diamr, 20-In stroke, 80 Strokes pr Minute, and considering the great number of air leaks from such extent of pipe and number of machines, must reduce the pressure of the Atmosphire on each piston to a verry small weigh, unless this air-pump and steam engine is of an immence size beyond all bounds. These objections I made them aquainted with, and told them that, before they went to such an expence, that it would be a safe plan to first make further inquirey, respecting the probility of the plan answering and espeshally to make the first experiment on a shure plan, becase all the other docks Compainys are looking to the result of this experiment. At the time I was inform'd of this plan, a thought struck me that it might be accomplish'd by another mode preferable to this: by having one powerfull steam-engine to force water in pipes round the dock, say 30 or 40 pounds to the inch, more or less, and to have a worm shaft, working in to a worm wheel, exactley the same as a common roasting jack, and apply to the worm shaft a spouting arm like barker's mill horizontally, and the worm shaft standing perpenduclear would work the worm wheel thats on the chain barrell shaft of the crane, which would make the machine very simple and cheap, and accomplish a circulear motain at once, instead of a piston alternitive motain to drive rotarey motain. This report of mine had som weight with them; and an arrangement is on foot to make inquirey in to the plan propos'd by me, so as to remunerate me, provided my plan is good and Mr Linthorn wishes an investigation to take place before scintific and able judges, and requested me to name som one. I must beg your pardon for making so free as to request this favour from you by being for 30 years favoured with your advice, from which I am convinced that you are the most able man in england to give advice opponion on this plan. Mr Linthorn intends to request Dr Woolenston to accompainy you any day that it may be convenient, and in the mean time, should you see him, it might not be amiss to mention it to him; and should you be able to attend an hour or two to this buisness, I would thank you to drop me a not, saying when it may be convenient to you and where I shall meet you. There is a memorandum of an agreement between Mr Linthorn [and me], but the plan that I sujest is onely as yet made public to him and your self.

It should be remarked that Trevithick's ideas of hydraulic power transmission for working machinery had been anticipated by Joseph Bramah in his patent of 1795, and that the latter had actually reduced his system to practice in 1802. William Murdock had distributed compressed air to a limited extent at Soho early in the century, but both methods of power transmission had to wait for many years of hard work before becoming practicable.

Trevithick still overflowed with ideas, as the next letter shows:[1]

Fancy and whim still promps me to trouble you, and perhaps may continue so to do untill I exhaust your patience. A few days since I was in company where a person who was saying that as much as one hundred thousands per year was paid in this place for the use of ice, the greatest part of which was brought by ships sent to the greenland seas for that express purpose. A thought struck me at the moment that artificial cold might be made very cheap by the power of steamengines, by compressing air in to a condencer surrounded by water, and also an injection in to the same, so as to instantly could down the verry high compress air to the tempture of the surrounding air, and then admitting it to escape into liquid. This would reduce the tempture to any state of cold required.

---

[1] *Life*, II, 294. Letter to Giddy, dated 1828, June 29th, address as previously.

He concludes:

However, my information on this subject is not sufficient to satisfy my own mind, and I take the liberty to trouble a much more fertile brain than my own, in hopes to draw a satisfactory conclusion before [I] drop the idea.

This was a pregnant idea, but only an idea, as a glance at the difficulties that had to be encountered before it became an accomplished success, some forty years later, show.

Trevithick went to Holland, in what capacity we do not know, to report on the applicability of the Cornish engine for drainage purposes. In connection with this visit his son relates a story so typical of Trevithick's good nature that we cannot refrain from quoting it:[1]

He had not money enough for the journey, and borrowed 2l. from a neighbour and relative, Mr John Tyack. During his walk home a begging man said to him, "Please your honour, my pig is dead; help a poor man". Trevithick gave him 5s. out of the 40s. he had just begged for himself. How he managed to reach Holland his family never knew; but on his return he related the honour done him by the King at sundry interviews, and the kindness of men of influence in friendly communion and feasting.

We gather that he left England about July 16th or 17th and spent ten days in Holland; one cannot imagine that £2, even if he had not previously parted with any of it, would be of much use for such a journey!

What he did and saw is related in a letter to Giddy which exhibits such a grasp of the situation and a practical way to cope with it, that we give it in full:[2]

The night before last I arrived from Holland, where I spent 10 days. I found my relative there, Nichs Harvey,[3] the son of John & Nancy Harvey. He is the engineer to the Steam Navigation Company at Rotterdam. They have a ship 235 feet long of 1500 Tons burthen, with

---

[1] *Life*, ii, 295.

[2] *Enys Papers. Life*, ii, 295. Letter dated London, 1828, July 31st.

[3] Nicholas Oliver Harvey (1801–1861) was in high favour in Holland, see *Trans. Inst. C.E.* xxi, 558, and doubtless the introduction came about through him.

three 50 Inch Cylinders double, also two other ships 150 feet long, with each two double 50 Inch Cylinders, on board ready to take troops for Batavia. The large ship and three engines cost eighty thousand pounds. The Steam Navigation Company built them, and also a great many others of different sizes. This Company have been long anxious to get me to Holland, having heard of the duty preformed by the Cornish engines, and was anxious to know what might be don towards draining and relieveing Holland from the present ruinous state. Immidly on my arrival there I joined the Dutch Company, and entered into bonds with them. I give you, as near as I can, the state of the Country at present. Aboute 250 years since, a strong wind threw a bank of sand in the mouth of the river Rhine, which made it overflow its banks; when eighty thousand lives was lost, and above forty thousand acres of land, which remains to this time under 12 feet of water. About 100 years since the bead and surface of the river Rhine was 5 feet below what it now is, and the under floors of the houses in Holland is nearly useless, and in another centry must be totaly lost, unless something to prevent it is don, as the river at present is nearley overflowing its banks, and in consiquence of the rise of water, the windmill engines cannot lift oute the water, and to erect steam-engines, they never could beleive would pay the expenses. Nearley one half of Holland at present is under water, either totaly or parshally, becase the ground that's keepd dry in winter is flooded in summer for want of wind to lift it oute. Therefore their crops of corn [are] often totaly lost. Aboute 6 years since it was in contemptlation to recover the forty thousand acres before stated, and a Company was made up of the King and principal men in Holand, to drain this by windmills, which they estamated would cost two hundred and fifty thousand pounds, and that making the banks and canals four hundred and fifty thousand more, when made by men's labour, and 7 years to accomplish it. This 7 years was a great objection, becase of the unhealthey state of the Country while draining. The water is aboute 18 inches every year to be lifted, on an average abt 10 ft high. I have been furnished with very correct calculations and drawings from this Company, who expected to have drained forty thousand acres in 7 years, at an expense of seven hundred thousand pounds, which, when drained, would have sold at £50 per acre, aboute two millions. I fiend, that from the statement given me, of 18 Ins of water to be lifted 10 feet high, that it would require aboute one bushell of coals to lift the water from one acre of ground for one year, and that a 63 In Cylinder double would perform this work from 40,000 acres, when working with high steam and condencing, at an expense of less than three thousand [pounds] per year; and engines in boats would cut and make the inbankments and canals, withoute the help of men. I propos'd six Cylinders of 60 Ins Diamr, Double power, which would drain the water in one year; and also 4 others for cutting the canals and making inbankments. The whole expenses would not exceed one hundred thousand pounds and one year, instead of seven hundred thousand pounds and seven years; and above sixty thousand acres more [are] to be drained in other spots, besides all the summer's water to be lifted, from the ground that is now partically drained. Farther it was proposed by Goverment to cutt open the river Rhine to one thousand yards wide and 6 feet deepe for 50 or 60 miles in length; a sum they supposed would cost them ten millions sterling. I have proposd to make Iron ships of one thousand ton burthen, with an engine in each, which would load them, propell, and also empty them for about one penny per ton. Each ton will be aboute one square yard, and the cutting the river Rhine to one thousand yards wide, 6 feet deeper, for 50 or 60 miles in length, will not cost one and a half millions and be accomplished in a short time. I farther propos'd that all this rubbish be carried in to the sea of the Uder [i.e. Zuyder] Zee, which would make dry by imbanking it with the rubbish, near one Million acres of good land, capable of paying ten times the sum of cutting open the river Rhine, which all togeather would add one hundred per Cent. more to the surface of Holland, and at this time its much wanted, becase their settlements abroad is fell allmost to nothing, and they have much to many inhabatants for the land at present. I made them plans for carrying the whole in to effect, and I have clos'd my agreement with them; and in a few days I shall go down to Cornwall, and promis'd to return again to Holland within one month from the time I left it. I saw Mr Hall and the Engineer of the Dock Compy to-day, who are satisfyde that the plan for working the cranes is a good one. I am to see them again on monday next; after which I shall return home, where I hope to see you, to consult you on the best plan to construct the machinery for heaving the water, and cutting the canals, and making the dykes.

Francis Trevithick says:[1]

A drainage company was formed in London with a board of directors, some of whom thought that a new kind of engine should be invented and patented as a means of excluding others from carrying on similar but competing operations. Trevithick, always ready to invent new things, though never forgetting his experience with old things, instinctively returned to the Dolcoath engines, and recom-

[1] *Life*, II, 297.

mended them as suitable for the pumping work; but finally a new design was determined on, and Harvey and Co., of Hayle, received orders for the construction, with the greatest possible dispatch, of a pumping engine for Holland.

As the lift was very low, Trevithick adopted the rag and chain pump, but instead of the usual discs he used balls. It was of large size, for the diameter of the balls, i.e. of the pump barrel, was 3 ft., while the steam cylinder was 3 ft. diam. and 8 ft. stroke. The whole was arranged with boiler self-contained in an iron barge. The pump was, as would be expected, noisy, but this was partly cured by substituting links between the balls instead of the chain. Trevithick's own description is as follows:[1]

The first engine that will be finished here for Holland will be a 36 Ins Cylinder and a 36 Ins water pump, to lift water abt 8 feet high; on the crank-shaft there is a rag head of 8 ft Diamr, going 8 feet per Second, with balls of 3 ft Diamr passing through the water pump, which will lift about 100 tons of water per minute. Its in a boat of iron, 14 ft wide, 25 ft long, 6 ft high, so as to be portable, and pass from one spot to another, withoute loss of time. This will drain 18 Ins deep of water (the anual produce on the surface of each acre of land) in aboute 20 minutes to drain each acre, with aboute one Bushell or 6d worth of Coals per year. The engine is high pressure and condencing.

In another letter to Giddy on January 24th, 1829, Trevithick says:"The engine for draining in Holland will be ready to send there aboute the end of Febry". Apparently it was not completed quite so soon, for in a letter to the same, July 27th, Trevithick writes:[2] "The Holland engine lifted on tryl, when the[y] came down to inspect it, 7200 Gallns of water per minute 10 feet high with one bushell of coals per hour—exceeding good duty for a small engine of 24 Ins Cylinder 34,560,000 Duty".

By the way, the cylinder was stated previously to be 3 ft. diam. This is the largest chain pump of which the authors have

[1] *Enys Papers. Life*, II, 317. Letter to Giddy, dated Hayle Foundry, 1828, Dec. 14th.

[2] *Enys Papers. Life*, II, 333. Letter dated Hayle Foundry, 1829, July 27th.

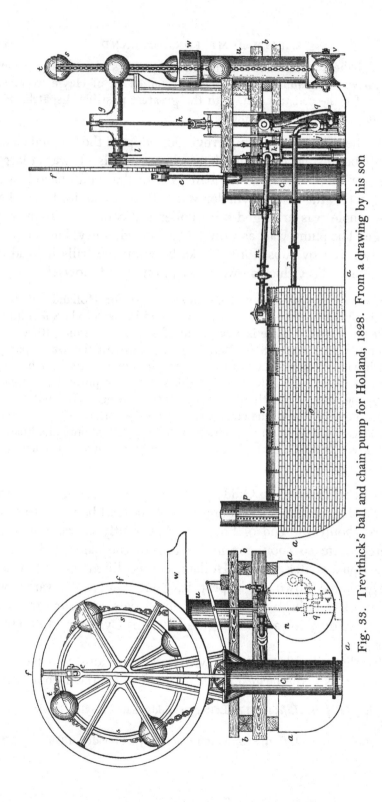

Fig. 33. Trevithick's ball and chain pump for Holland, 1828. From a drawing by his son

any record. The engine, however, never went to Holland. Francis Trevithick, who was to have erected it, says:[1]

The directors having desired the writer to take the engine to Holland and set it to work with the least possible delay, adjourned for refreshment before starting for London. In those few minutes differences arose, resulting in the engine remaining for months in the barge, and then going to the scrap heap.

It has been said above that Trevithick's hopes of getting any reward for his inventions from the mining interests were soon dashed. He quickly dropped the idea in favour of a petition to Parliament. He had plenty of precedents, for excluding grants such as that to Dr Edward Jenner for vaccination as not being comparable with a mechanical invention, there had been a grant to Edmund Cartwright of £10,000 in 1809 for the power loom, one to Samuel Crompton of £5000 in 1812 for the spinning mule, and one to John Palmer of £50,000 in 1813 for stage coaches. The prospect seemed therefore quite hopeful.

Trevithick mentions his intention in a letter[2] to Mr Edmonds, where he says that he was at Dolcoath account on December 17th "and made known to them my intention of applying to Government and not to individuals for remuneration. They are ready to put their signatures to the petition, and so will all the county. I fear that it is as much as we shall do to get it before the House in time". Like so many persons, the adventurers were ready to put their hands to a petition rather than into their pockets for ready cash.

Mr Edmonds drew up the Memorial, and as it gives such a succinct account of Trevithick's life's work we venture to recite it in full:[3]

To the Honourable the Commons of the Kingdom of Great Britain and Ireland in Parliament Assembled.

The Humble Petition of Richard Trevithick, of the Parish of Saint Erth, in the County of Cornwall, Civil Engineer, 27th February, 1828,

SHEWETH:

That this kingdom is indebted to your petitioner for some of the

[1] *Life*, ii, 301.
[2] *Life*, ii, 307. Letter dated 1827, Dec. 20th.
[3] *Life*, ii, 307.

most important improvements that have been made in the steam-engine, for which your petitioner has not hitherto been remunerated, and for which he has no prospect of being ever remunerated except through the assistance of your Honourable House.

That the duty performed by Messrs Boulton and Watt's improved steam-engines in 1798, as appears by a statement made by Davies Gilbert, Esq., and other gentlemen associated for that purpose, averaged only fourteen millions and half (pounds of water lifted 1 foot high by 1 bushel of coals), although a chosen engine of theirs, under the most favourable circumstances, at Herland Mine lifted twenty-seven millions, which was the greatest duty ever performed till your petitioner's improvements were adopted, since which the greatest duty has been sixty-seven millions, being more than double the former duty. That prior to the invention of your petitioner's boiler the most striking defect observable in every steam-engine was in the form of the boiler, which in shape resembled a tilted waggon, the fire applied under it, and the whole surrounded with mason-work. That such shaped boilers were incapable of supporting steam of a high pressure, and did not admit so much of the water to the action of the fire as your petitioner's boiler does, and were also in other respects attended with many disadvantages.

That your petitioner, who had been for many years employed in making steam-engines on the principle of Boulton and Watt, and had made considerable improvements in their machinery, directed his attention principally to the invention of a boiler which should be free from these disadvantages; and after having devoted much of his time and spent nearly all his property in the attainment of this object, he at length succeeded in inventing and perfecting that which has since been generally adopted throughout the kingdom.

That your petitioner's invention consists principally in introducing the fire into the midst of the boiler, and in making the boiler of a cylindrical form, which is the form best adapted for sustaining the pressure of high steam.

That the following very important advantages are derived from this, your petitioner's, invention. This boiler does not require half of the materials, nor does it occupy half the space required for any other boiler. No mason-work is necessary to encircle the boiler. Accidents by fire can never occur, as the fire is entirely surrounded by water, and greater duty can be performed by an engine with this boiler, with less than half the fuel, than has ever been accomplished by any engine without it. These great advantages render this small and portable boiler not only superior to all others used in mining and manufacturing, but likewise is the only one which can be used with success in steam-vessels or steam-engine carriages. The boilers in use prior to your petitioner's invention could never with any degree of safety or convenience be used for steam navigation, because they required a protection of brick and mason work around them, to confine the fire by which they were encircled, and it would have been impossible, independent of the great additional bulk and weight, that boilers thus constructed could withstand the rolling of vessels in heavy seas; and notwithstanding every precaution the danger of the fire bursting through the brick and mason work could never be effectually guarded against.

That had it not been for this, your petitioner's, invention, those vast improvements which have been made in the use of steam could not have taken place, inasmuch as none of the old boilers could have withstood a pressure of above 6 lbs. to the inch, much less a pressure of 60 lbs. to the inch, or even of above 150 lbs. to the inch when necessary.

That as soon as your petitioner had brought his invention into general use in Cornwall, and had proved to the public its immense utility, he was obliged in 1816 to leave England for South America to superintend extensive silver mines in Peru, from whence he did not return until October last. That at the time of your petitioner's departure the old boilers were falling rapidly into disuse, and when he returned he found they had been generally replaced by those of his invention, and that the saving of coals occasioned thereby during that period amounted in Cornwall alone to above 500,000*l*.

That the engines in Cornwall, in which county the steam-engines used are more powerful than those used in any other part of the kingdom, have now your petitioner's improved boilers, and it appears from the monthly reports that these engines, which in 1798 averaged only fourteen and half millions now average three times that duty with the same quantity of coals, making a saving to Cornwall alone of 2,781,264 bushels of coals, or about 100,000*l*. per annum. And the engines at the Consolidated Mines in November, 1827, performed sixty-seven millions, being forty millions more than had been performed by Boulton and Watt's chosen engine at Herland, as before stated.

That had it not been for your petitioner's invention, the greater number of the Cornish mines, which produce nearly 2,000,000*l*. per annum, must have been abandoned in consequence of the enormous expense attendant on the engines previously in use.

That your petitioner has also invented the iron stowage water-tanks and iron buoys now in general use in His Majesty's navy, and with merchant's ships.

15-2

That twenty years ago your petitioner likewise invented the steam-carriage, and carried it into general use on iron rail-roads.

That your petitioner is the inventor of high-pressure steam-engines, and also of water-pressure engines now in general use.

That his high-pressure steam-engines work without condensing water, an improvement essentially necessary to portable steam-engines, and where condensing water cannot be procured.

That all the inventions above alluded to have proved of immense national utility, but your petitioner has not been reimbursed the money he has expended in perfecting his inventions. That your petitioner has a wife and large family who are not provided for.

That Parliament granted to Messrs Boulton and Watt, after the expiration of their patent for fourteen years, an extension of their privileges as patentees for an additional period, whereby they gained, as your petitioner has been informed, above 200,000*l*.

That your petitioner therefore trusts that these his own important inventions and improvements will not be suffered to go unrewarded by the English nation, particularly as he has hitherto received no compensation for the loss himself and his family have sustained by his having thus consumed his property for the public benefit.

Your petitioner therefore most humbly prays that your Honourable House will be pleased to take his case into consideration, and to grant him such remuneration or relief as to your Honourable House shall seem meet.

And your petitioner, as in duty bound, will ever pray, &c.

The petition was put into the hands of Giddy, now M.P. for Bodmin, for presentation to Parliament. What exactly happened to it we do not know; all we learn about it is, in a letter to Giddy, that it was unsuccessful:[1] "I find that Mr Spring Rice cannot get the Lords of the Treasury to agree to remunerate or assist me in any way".

Of course they could not do so without the sanction of Parliament, so that there must have been some misdirection of activities; yet it is difficult to believe that when Trevithick had behind him an M.P. who "knew the ropes" like his friend Giddy, any such misdirection had occurred. Probably the matter had not been pushed in the right way or insistently enough, for although Parliament was rather averse to making such grants,

---

[1] *Life*, II, 311. Letter written from Lauderdale House, 1831, Dec. 24th.

taking the view generally that the Patent Law offered a remedy, yet John Loudon McAdam in 1827 received a gratuity of £2000 and a sum of £8000 for past services for his improvements in road-making, and Morton in 1832 got a grant for his slipway for repairing ships. No such grant was made to Trevithick and thus another of his hopes was dashed to the ground.

Trevithick had, as we have seen from the Tincroft days, taken the keenest interest in the performances of engines in Cornwall. Sad to relate, the performances since Boulton and Watt left Cornwall had retrograded, a situation that was viewed with concern by those who had the mining interests at heart. Hence a proposal that reports of engine performances should be published regularly was advocated; Captain Joel Lean for many years was foremost in doing so. Just prior to his return to Cornwall after his severe illness, Trevithick had been approached to see if he would undertake such a part-time job. The enquiry came from his brother-in-law to Mrs Trevithick:[1]

I saw Captain Andrew Vivian on Wednesday, who told me that he had been offered 150*l.* a year to inspect all the engines in the county, and report what duty they were doing, in order to stimulate the engineers. He declined accepting it, having too much to do already; and he thought it would be worth Trevithick's notice, as it would not take him more than a day or two in a month.

The job was obviously a routine one unsuited to his temperament and, rightly, he did not accept it. Happily, Captain Joel Lean himself was willing to undertake the task of "Registrar and Reporter", work which was carried on by his successors, his sons Thomas and John. Reporting started in 1811 with twelve engines, and the rivalry stimulated among engineers by the publication of their duties led to steady improvement. The duty of 17 millions in 1811 rose by 1816 to 23 millions, the average of thirty-five engines.[2]

In the year 1826, just prior to Trevithick's return, the number

---

[1] *Life*, II, 154. Letter dated Hayle Foundry, 1810, Aug. 26th.

[2] Cf. Lean, Thomas and Brother [John], *Historical Statement of the* ... *duty performed by the Steam Engines in Cornwall,* 1839.

of engines reported was fifty-one and the average duty 30·5 millions. Such improvement, which is progressively more difficult, as we all know, had been made possible by the fact that since 1822 "many mines replaced their machinery by new and improved engines; and considerable excitement and emulation appeared among the engineers".[1]

In 1827 the number of engines remained at fifty-one and the duty was slightly improved, viz. 32·1 millions, but what made this year so memorable a one in the history of the Cornish engine was the trial of Captain Samuel Grose's 80 in. engine at Wheal Towan, which attained a duty of 62·2 millions, and Arthur Woolf's new 90 in. engine at the Consolidated Mines, which attained the unprecedented performance of 67 millions. Distrust and doubt were freely expressed, and such excitement prevailed that a public trial of the last-named engine was decided upon. The report, dated December 19th, 1827, was that the duty attained was 63·6 millions. Excitement was further increased when Captain Grose, by the introduction of several improvements at Wheal Towan, went one better and in April, 1828, actually reached 87 millions; this duty was confirmed by another public trial reported on May 7th. Trevithick's friend, Giddy, had also taken up the matter of "duty" and had read a paper[2] before the Royal Society entitled "Observations on the Steam Engine", in which he gave a résumé of the existing conditions in Cornwall.

Such was the excitement and stir which Trevithick found on his return, but in which he had no part or lot, although Captain Grose had been a pupil and Woolf a former employé of his. Besides, Trevithick may have felt that the efforts of the engineers were somewhat belated, for had he not obtained a duty of 28 millions, non-condensing, at Herland Mine in 1815?[3]

In August, 1828, Trevithick, accompanied by his son Francis, who seems to have been at this period a constant companion both in the field and in the drawing office, went round the mines

---

[1] *Loc. cit.* p. 56.
[2] *Phil. Trans.* Jan. 25th, 1827.
[3] *Life*, II, 325.

collecting data about the performance of engines, and a note-book was in existence in which he had recorded particulars of engines at Wheal Towan (already mentioned), Wheal Vor, Wheal Damsel and Binner Downs. In the last-mentioned case, the cylinder and steam pipes were encased in brick-work, leaving an annular space through which hot gases could circulate from a separate fire. When the latter was put on, the super-heating caused thereby increased the duty from 41 to 63 millions with an added expenditure of fuel of only 5 bushels of coal per day.

Trevithick communicated his observations and reflections on this subject to Giddy in a letter which, though long, must be quoted:[1]

On my return from London 5 Weeks since I was dissappointed at not finding you in Cornwall. I have made enquirey into the duty performd by the best engines, and the circumstances they are under, from which it appears to me to be som verry unacountable circumstances which as yet have not been accounted for and particulear in Binner Downs Engines. The statement given to me was from Capn Gregor, the cheaf agent and engineere of that mine, which appeared to me to be so plain that I cannot doubt their facts, tho' they differ so very widely from all former opponions. In this mine there are two engines, one of 42 Ins Diamr, the other of 70 Ins Diameter, 10 ft stroke. Formerly those engines workd withoute Cylinder cases, at which time the 70 Ins Cylinder burned 1½ wey of coals and did regulier a duty of 41 Mills; since which brickwork have been placed round the cylinder and steam pipes, leveing a narrow flue aboute the engine, which is heated by seperate fires. These fires consumes abt 5 bushells of coals in 24 hours; the heat aboute the Cylinder is not so great as to injure the packing which stands good for 13 week and the hand can be sufferd in the flue; the saving is such as to make now for several months past 63 millions. Before this flue was placed the Coals consumd under the boiler was 108 Bushells, now onley 67 Bushells consumd, under the boiler which with the additional 5 bushels burnt round the flue makes it 72 Bush of Coals. Therefore for the Coals burnt under the boiler a duty of 66 Millions is performd, which makes an expand-tion of above 60 per cent. occasioned by the heat of the flues by 5 bushells of coals, and a duty of 1781 Millions gained in 24 hours by

1 *Enys Papers*. *Life*, II, 315. Letter dated Hayle Foundry, 1828, Dec. 14th.

5 bushells of coals, which amounts to 350 millions gained by each of these 5 bushells. The 42 Ins Cylinder is as near as possible under the same circumstances, no other alteration what ever have been made; and by way of a proof of this, they have left oute the fires round the flues, and immidly the engines falls back to their former duty, and the condencing water increase in the same proportion. The Steam pipes and cylinder surface sides thats heated by this 5 bushells of coals amounts to aboute 300 Surface feet, and the saving amounts to 1781 millions, which is 6000 mills saving for each of the 300 feet of surface on the castings. In Whl Towan engine that did 87 Mills, the surface sides of the boiler was 1000 feet of fire side for every Bushell of coals burnt within the hour, and the duty performd per minute from each foot of boiler firesides was 1500 pounds one ft high. Now it appears that the heating of Binner Downs 300 feet of surface feet gave a saving of 6000 pounds per mint for each foot of surface; whereas the boiler sides onely gave 1500 pounds of duty pr mt for each foot of firesides in the boiler. Therefore the savings on heating the sides of the Cylinder is equal to four times the duty don by each square foot of boiler sides; and farther, it appears that the 300 feet of surface, when not heated externelly, tho' clothed round with brick work, condenced or prevented from expanding the steam of 41 Bushells of coals, which was 8 times as much Steam condenced as 5 bushells of coal would rise. Now if this report be facts, which I have no reason to doubt (but still I will be an eye witness to it next week,) there must be an unknown propensity in steam above atmosphire strong to a very sudant condencation, and *vice versa*, to also a suddant expanshon, by a verry small heat applied to the steam sides; and if by heating steam, independant of water, such a rapid expanshon takes place with so little heat, certainly a rapid condenceation must take place in the same ratio, which might be don at sea by cold sides to a great advantage, always working with fresh water. I shall have a small portable engine finishd here next week, and will try to heat steam, independant of water, in small tubes of Iron, in its passage from the boiler to the Cylinder, and also try cold sides for condencing. If the before statements should prove to be correct, allmost everything might be don by steam, becase then water would not be wanted for portable engines, but parshially condencd and again returned into the boiler, withoute a fresh supply or incumbred with a great quantity; and boilers might be made of verry extencive fire side, both to heat water and also steam, and yet be verry light. It appears that this engine, when working withoute the heated flues round the Cylinder & pipes, avaporated 20,000 Gallns of water more into steam, in 24 hours, than when heated, and the increase of con-

dencing water required in the same propotian. Its so unaccountable to me that I shall not be satisfyed untill I prove the fact, the result of which I will inform you. I should be verry glad to receive your remarks on the foregoing statements.

Giddy replied by sending him a copy of his paper "Observations on the Steam Engine"[1] already alluded to.

Trevithick answered it in a still longer letter than before; again it is best to give it very fully:[2]

On the 28th inst. I recd enclosed your printed report on Steam, and I have also examined Farey's publication on sundry experiments made by Mr Watt, which are very far from agreeing with the actual performance of the Engines at Binner Downs. Mr Watt says that steam at one atmosphere expands 1700 times its own bulk of water at 212 Deg, with 15 lb to the inch pressure, and that large Engines ought to perform 18 millions when loaded with 10 lb. to the inch actual work, the condensing water being one fortieth part of the content of the steam of atmosphere strong in the Cylinder, cold water at 50 Deg, and hot water at 100 Deg, which would be for a 70 Inch Cylinder such as Binner Downs of a 10 feet stroke, 11 lb. to the inch, which is one tenth more load than Mr Watt's, whos calculation for that size Engine, stroke and load according to his tables laid down by Farey would have been 57 Gallons of injection water for each Stroke, heated from cold water of 50 to hot water of 100 Degs and to do 18 Millions would consume, when working 8 strokes per minute, 11¼ Bushells of Coals per hour. Now the actual fact by proof of Binner Downs Engine, is that she worked 8 strokes per minute 10 feet stroke 11 lb. to the inch with 3 bushells of coals per hour, and with 13 gallons of injection water to each stroke heated from cold water of 70 Deg to hot water of 104 Deg, which is a heat of 34 Deg above the heat of the mine water multiplied by 13 gallons, makes 442: and by Mr Watt's tables their Engine doing the same work, it would require 57 Gallons condensing water heated from 50 Deg, to 100 Deg, 50 Deg which would have been raised, multiplied by 57 Gallons [amounts to] 2850, which is 6½ times the quantity of condensing water and near 4 times the quantity of coals that is actually made use of at present.

Trevithick goes on to make some calculations whereby he reaches the conclusion:

[1] *Phil. Trans.* Jan. 25th, 1827.
[2] *Life*, II, 323. Autograph letter dated Hayle, 1828, Dec. 30th.

...therefore, by Mr Watt's view it appears that low steam would do one-fifth more duty than high steam, and while Binner Downs engine by actual work performs about four times the duty of Mr Watt's theory and practice, and only one sixth part of the heat carried of by the condensing water, which proves that high steam have much less heat, for its effective force; and farther proved by the small quantity of condensing water its heat is extracted with much less cold water. Yesterday I proved this 70 Inch Cylinder while working with the fire round the Cylinder, which flues only consumed 5 bushells of coals in 24 hours. The Engine worked 8 strokes a minute, 10 feet stroke, 11 lb. to the inch; steam in the Boilers 45 lb. above the atmosphere, with 12 Bushells of coals for 4 hours, and 13 Gallons of condensing water to each stroke, heated from 70 Deg to 104 Deg; but when these fires round the cylinder was put out, though still having a coating of hot brick work round the Cylinder, and performing exactly the same work, yet it burnt 17 Bushells of coals in 4 hours, and required 15½ Gallons of condensing water, heated from 70 Deg up to 112 Deg. Therefore you will find that the increase of the consumption of coal, by taking away the fire from round the Cylinder, was very nearly in the same proportion as the increase and temperature of the condensing water, which shows the experiment [to be] nearly correct.

Trevithick proceeds to adduce "a further proof of the more easy condensation of high steam" by giving the results of experiments on Binner Downs 42-inch engine and arrives at this general conclusion:

Its my opinion that high steam will partially expand and contract with a much less degree of heat or cold for its effect, than what steam of atmosphere strong will do, and I intend to try steam of 5 or 6 atmospheres, and partially condense it down to nearly atmosphere strong, and then by having an air-pump of more content than is usual to return the steam, air, and water, from the top of the air-pump, all back into the boiler again, above water-level in the boiler, and by having a great number of small tubes, with great hot surface sides, to reheat the returned steam (without the half of water in contact with it); though by this plan I shall lose the power of the vacuum, and also the additional power that is required on the air-bucket to force it back again into the boiler, yet by getting so much heat returned again, I think will over-balance the loss of power, besides having always a supply of fresh water, which in portable Engines, either on the water or on the road, would be of great value. I should esteem it a very great favour if you would be so good as to turn this over in your mind the probable theory of the remarks here stated, and give me your opinion. If Mr Watt's reports on his experiments is correct, how could it be possible that the result of the high-pressure engine that I built at the Herland 13 years since, which discharged the steam in open air, did 28 millions?

Trevithick offers to send Giddy a copy of the certificate of the Herland trial. He then puts a hypothetical case of a steam engine worked on a cycle like a hot-air engine and concludes: "This mystory ought to be laid open by experiments, for what I have stated are plain facts from actual proofs, and from which I have no doubt but that time will show that the theory of Mr Watt is incorrect".

It must be realized that the properties of steam were not fully understood at the date of which we write. Watt had carried out experiments on steam in 1771 and 1783; John Southern and William Creighton, his talented assistants, had repeated them more accurately and extended them in 1814, but they were not published till 1818, in the *Encyclopaedia Britannica*.[1]

The important observation of Trevithick that high-pressure steam will expand or contract with a much less degree of heat or cold than will steam of, say, one atmosphere pressure was known and is succinctly expressed by Farey, the author quoted by Trevithick, in these words:[2] "The number of additional degrees of temperature which will produce an additional atmosphere of elasticity, is continually diminishing as the temperature and elasticity become greater". No one, however, had proposed to take any advantage of this property of steam. Trevithick entertained the idea, expressed shortly, of using steam in a closed cycle between high limits of temperature and pressure. As condensation was eliminated and, incidentally, the latent heat not thrown away, the scheme was applicable to locomotive engines. Trevithick in the next few months de-

---

[1] Articles "Steam" and "Steam Engines", by John Robison, LL.D., subsequently reprinted 1818, with additions. Later still this matter was embodied in Robison's *System of Mechanical Philosophy*, 1822, 5 vols.

[2] *Steam Engine*, 1827, p. 74.

signed and commenced the construction of an engine to embody his novel ideas. He describes what he is doing in letters to Giddy. In the first[1] he recounts the results of further experiments and his deductions from them such as:

I fiend that steam of 4 atmospheres being left to expand to its full extent onely takes cold water to condence in the same propotain as to the quantity or bulk of steam and not as the duty performed, and that no more heat is taken from the boiler in atmospheres than in one, tho' the performance is four times the effect.

In the further letter he says:[2]

Below you have a sketch of the engine that I am making here for the express purpose trying the experiment of workg the same steam and water over and over again, and heating the returned steam by passing [it] in small streams up thro' the hot water from the bottom of the boiler to the top. The boiler is 3 feet Diamr, stands perpendicular; with a tube for the fire in the middle of the same of 2 ft Diamr; and a steam-case round the oute side of the boiler 1½ Ins space. This case keeps the boiler hot, and at the same time parshally condenses the steam before its forced in to the boiler again. Boiler is 15 feet high; Cylinder 14 Inches Diamr, single power 6 ft stroke. The pump that forces the steam & water back again 10 Ins Diamr piston, 2 ft 9 In stroke, aboute one-quarter part of the content of the steam Cylinder. The bottom of the boiler will have a great number of verry small holes in it not above $\frac{1}{16}$th of an Inch Diameter, through which the steam deliver'd in to the boiler will pass up 15 ft through the hot water, by which I should think it will heat those verry small streams of steam again to their usual tempture. The pump for lifting cold water to prove the duty of the engine is 30 Ins Diamr, 64 ft Stroke, lengthen'd or shortened occasionally from 6 ft Lift to a 12 feet lift, as the tryal or load on experiment may require, which will be from 12 to 24 pounds to the inch on the steam piston. This machine will be ready for your inspection before you return to Cornwall, and I intend to proove it effectually before I go to Holland.

...What effect do you think that the water will have on heating the steam in its passage to the top of the water from the false bottom of the boiler? I have a Cistern of cold water, with a proper condencer in it, connected between the bottom of the boiler case and the force pump to the bottom of the boiler; therefore I can either parshally condence

[1] *Enys Papers.* Letter dated 1829, Feb. 6th.
[2] *Enys Papers. Life,* II, 332. Letter dated Hayle, 1829, July 27th.

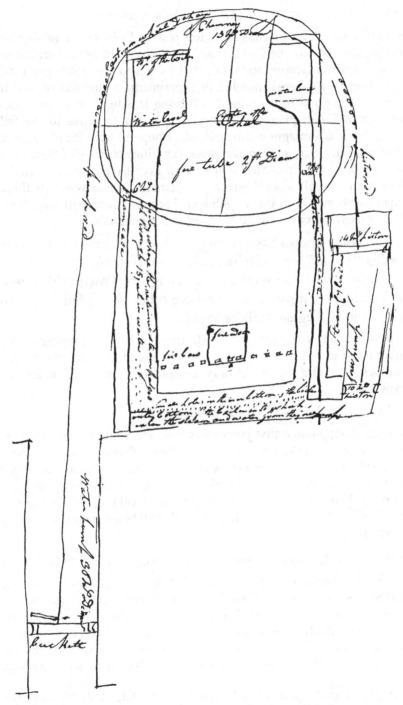

Fig. 34. Trevithick's closed cycle steam engine and boiler, 1829

by cold water sides or cold air sides, just as I please, by raising or sinking the water in the Cistern; and the boiler is made very strong to try different temptures, and additional length to the water pump all very suteable to a great number of experiments to any extent, and if there is any good in the thing I will bring it oute. I shall have indicators at different places and everything proper to prove to the full extent what advantages can be gained. I hope to have the pleasure of your company during to make these experiments, which I think will throw more light upon this subject than ever have been yet don.... Som farther tryals since I wrote you last still farther warrents these experiments and I am verry confident that much good will arise from these experiments,...but to what extent is uncertain.

The boiler shown has a strong resemblance to the one he made for his "recoil" or aeolipile engine (see p. 150).

Engine performance continued to engross Trevithick's attention almost exclusively and we have other letters full of it.[1] In the letter of August 13th he says:

If you could point oute any satisfactory method of attaining at any thing like a guideing theory, it might save me much time, trouble and expence in going the right way to construct this engine for the experiment.

In that of August 27th he says:

I am dissapointed that you cannot give me a lift towards grabling my road in the dark, and am rather surpris'd that after so much have been said and reserches made by so many men of science that as yet that no certain data have been given or even a trace to such that can be relay'd on. However, I will still push the tryal to a descive point by actual proofes, and have still hopes that I shall blunder onto something worth picking up.

Three months later the experimental engine was ready:[2]

The Engine have been Work'd. The result is 10 Strokes pr. mt., 6 ft Stroke, with a half a bushell of coals per hour, lifting six thousand pounds weight. This was don with water in the Cistern round the condencer, which water came up to 180 Degrs of heat, and remained to that pitch. The water side of the condencer covered with this hot water was 50 Surface feet of condencer....I tryd it to work with the

[1] *Enys Papers.* Letters to Giddy, dated 1829, Aug. 11th, 13th and 27th.
[2] *Life,* II, 335. Letter dated Hayle Foundry, 1829, Nov. 5th.

cold air sides, but I found that the cold air sides of 120 feet would not work it but 4 Strokes pr. minute. I should have Work'd the steam much higher than 50 lb. to the inch, but being an old boiler I thought it a risk. From what I can see I expect that each foot of cold air side in the condencer will do one thousand pounds one foot high per minute. I am now putting a boiler of 350 ft of cold sides more to the condencer, to give a fair tryal to condencing with cold sides alone. The steam below the piston was aboute 6 or 7 pounds to the inch above the atmosphire. The force pump to the boiler was aboute one fifth part of the Content of the Cylinder, and the valve close to the boiler lifted when the force piston was down aboute ⅔ of its Stroke, at which time the returned steam enter'd the boiler again. The Engine work'd very steady. . . . I have no doubt of doing near ten times the duty thats now don on board ships, with oute using salt water in the boilers as at present. Our boiler have been working 3 Days and the water have not sunk in it one inch per day. I am quite satisfyde that the tryal allready will warrant success.

Giddy was not able to go and see the new engine, but he showed his interest by correspondence. Meanwhile further trials had been carried out, as reported in Trevithick's next letter:[1]

I have rec^d your both letters and sketches which shall be put in hand immedley. I understand it perfectly well. Since I wrote you last, I have made several satisfactory tryals of the engine and do not think it necessary to make any farther experiments. The statement that I give you below may be depended on for a future data. The load of the engine was 6280 lb., being 20 lb. to the inch for a 20 In Cylinder work'd a 6 ft Stroke, 12 Strokes per minute, with ¾ of a bushell of coals per hour, which gave a duty of 36,172,800 for one bushell of coals, a duty far beyond any thing before don in this Country by a 20 In. Cylinder 6 ft Stroke; I believe nearly double what have been don for engines of this size and stroke before. The cold water sides round the condenser was 60 feet, and the water at 112 [degrees] Temp[era]ture, not having a sufficient stream of cold water to supply the cistern. Each foot of these cold water sides did 7536 lb. per minute, which is about three times the work don in this Country by the hot boiler sides; therefore the condencer need not be more than one third of the boiler sides, and by making the condencer of 4 In Copper tubes

[1] *Enys Papers. Life*, II, 336. Letter dated Hayle Foundry, 1829, Nov. 14th.

of $\frac{1}{32}$nd of an inch thick, it would stand in one twentyth part of the space of the boiler. I put a boiler naked long side the engine to try cold air sides; but it was verry rusty, and did not condence as fast as I expected. The engine work'd exceeding well, but slow. The duty performed for each foot of cold air sides was 565 lb. pr minute, aboute one-thirtenth part of the condencing of cold water sides, but we never vent[ure]d to get the steam above 60 lb. to the inch. I have no doubt but [that] copper pipes of $\frac{1}{32}$nd of an Inch thick, clean and small, would do considerably more, becase the hot water that came out of the boiler from the condenced steam was but 170 degrs, and the external sides the same heat when the steam was 15 lb. above the atmosphire in the condencing boiler. This boiler was 4 ft 6 In Diamr, and I think that towards the external sides of the boiler that there was a colder atmosphire, if I may call it so, than what it was in the middle of this large condencing boiler, becase I found by trying a small tin tube, that it would condence 1500 lb. for each foot of cold air sides. However, as it is, it will do exceeding well for portable purposes, and the duty, I have no doubt, will be, both for water and air sides condencing, at least 50 per Cent. above our Cornish engines, which will be above five times what is now don with ships' engines, and espeshally when you take into consideration their getting steam from salt water, and letting oute so much water from the boiler to prevent the salt from accumulating in the boiler, which will make thirty per cent. more in its favour. If strong boilers to stand 200 lb. to the inch are made with small tubes, I have no doubt but that the duty would be considerably more, and my engine will not be one quarter part of the weight, price or space of others; and I have no doubt when every advantage is taken it will be 1000 per cent. superior in savings of coals to those now at work on board. The engine works exceeding well and returns the steam very regular every stroke in to the boiler. I am extreamly sorry that you was not present to see these experiments made. Please to make your remarks on the statement sent you, with any further information that you may judge usefull. I shall now make drawings agreeable to my experiments for actual performance on board ships.

In order to make the trial it was necessary to obtain the use of some vessel temporarily. We imagine that Trevithick's attempts to get one through friends must have failed, for we learn nothing further till seven months had elapsed, when at the suggestion of Mr George Rennie he decided to try the Admiralty. The following is the letter he sent, after submitting the draft to both Rennie and Giddy:[1]

To the Right Honorable

The Lords Commissioners of the Admiralty,
&c., &c., &c.

MY LORDS

About one year since I had the honor of attending your honorable board, with proposed plans for the improvement of Steam Navigation; and, as you expressed a wish to see it accomplished, I immediately made an Engine of considerable power for the express purpose of proving by practice, what I then advanced in theory; the result of which has fully answered my expectations: in consequence of this, I now make the following propositions to your honorable Board; that this entirely new principal and new mode, may be fully demonstrated for the use of the Public, on a sufficient scale. I humbly request your Lordships will grant me the Loan of a Vessel, of about two or three hundred Tons burthen for that purpose; in which I will fix, at my own expense and risk, an Engine of suitable power to propel the same, at the speed required: no alterations whatever in the vessel will be necessary; and the whole of the Apparatus required to receive its propelling force from the water can be removed and replaced again, with the same facility as the sails are managed, leaving the Ship without any apparatus whatever, beyond its sides when propelled by the wind alone: and, when it is propelled by steam alone, the apparatus outside of the Ship, will scarcely receive any shock or inconvenience from a heavy sea. This new invention entirely removes the great objection of feeding the Boiler with salt and foul water, none being required; and not one sixth part of the fuel, room, and weight of machinery will be required as is now in practice; it is also much more simple and safer not only for the purpose of navigation but for all other purposes where locomotive power is required, and will supercede all animal power; as the difficulty of getting and carrying water in Locomotive engines, is entirely removed: it will therefore prove on investigation, of greater utility than anything that was ever introduced to the Public.

I have to beg the great favour of your Lordships to appoint, not only scientific but practical Engineers to inspect my Plans; that you may be

---

[1] Record Office, Admiralty Sec. In Letters, 4540. Autograph Letter.

perfectly satisfied of the utility of them; not only as to the theory, but also to the practicability of carrying the same into full effect.

I have the honor to be

Your Lordship's most devoted, humble Servant

(Signed) RICHARD TREVITHICK.

Lauderdale House,
Highgate, June 15th 1830.

With it is the following holograph letter, dated June 17th, from Giddy to Mr (afterwards Sir) John Barrow, Secretary to the Admiralty:

My extraordinary countryman Trevithick has sent the enclosed letter under a cover to me, sealed as it is and without informing me of its contents, and desiring that I would forward it. I presume that it relates to steam and knowing Trevithick to be one of the most ingenious men in the world, although he has never done anything for himself, I have complied with his rather singular request.

The letter is minuted on the 18th:

Desire N[avy] B[oard] to cause a proper Person to inspect the Engine & Plan of this Person & to report their opinion to their L'd'ps as to how far they consider it advisable to entertain the proposal by allowing it to be tried on board one of H.M. Vessels.

Trevithick wrote again on July 19th, and this letter was also referred to the Navy Board. The officer directed to examine the matter was his old friend Goodrich. He called on Trevithick at Highgate but was far from being favourably impressed with the idea. He reported: "I still doubt one part of his theory regarding the contraction of the steam by cold water applied externally without a proportionate part of hot steam being reduced to the state of water, on which he grounds a great saving from not loosing as he considers the latent heat". Goodrich concludes his report: "I leave to the Board's consideration the question whether they would deem it expedient to recommend the loan of a Vessel to Mr Trevithick". The scheme being thus effectually damned by Goodrich's faint praise, it occasions no surprise to learn that the application hung fire.

The kind of boiler that Trevithick was experimenting with was a vertical tubular one with an annular casing through tubes in which the air for combustion is drawn (see Fig. 35). Trevithick's letter to Gilbert is somewhat striking, as it showed that he was an apt pupil of his master:[1]

The design is for the cold air to pass down from the top of the boiler through the air tubes within the steam case that's round the boiler, and to get heated in its passage by condencing the steam in the case, and to pass up through the fire bars in the hot state, nearly as hot as steam; becase as this air will be heated to nearly 212 [degrees] by condencing the steam in its passage withoute any of its oxigan being burnt, it will not carry off so much heat from the fire as cold air would, and still have the same oxigan as cold air to consume the coals. Therefore the cold air will be passing down through the steam case in the air tubes, and passing up through the fire and fire tubes in the boiler. I fiend by experiments that I have made here, that by putting a 2½ Ins Diameter tube of 4 ft long, in to a 4 In tube of the same length, with boiling water and steam between the both, keep'd up by a fire round the oute side tube, with a smith's bellows blowing in at the bottom of the inside tube, that this inside tube having 2⅔ Surface feet of condencing sides, on its inside, where the air is passing up from the bellows, do heat from 60 Degrs. to 134 Degrs. 15 Square feet of cold air per minute.

He goes on to give long deductions from these data and says:

I am anxious to have your opponion on this plan of returning the hot air from the condencer in to the fire place, and what you think the effect. The Comptroler of the Navy is not yet return'd from Plymouth, therefore no answer have yet been given to me.
...As I am verry anxious that every possible improvement should be considered prior to the specifycation for the patent made out, I must beg that you will have the goodness to consider and calculate on the data I have given you.

He proceeded to take out a patent, February 21st, 1831, No. 6082. No drawing was enrolled, but by the description it is clear that the tubular boiler, condenser and air vessel were formed of six concentric vertical tubes respectively. The inside tube comprised the fireplace and flue; the next annular space was for water and steam; the third annular space was filled with a

[1] *Enys Papers. Life*, II, 339. Letter dated Lauderdale House, 1830, Aug. 20th.

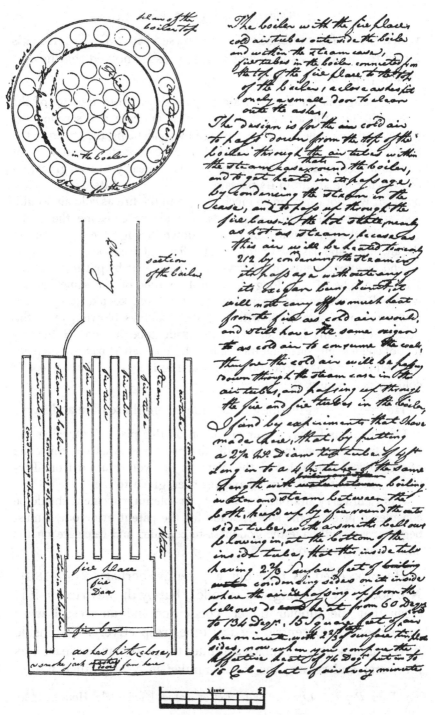

Fig. 35. Trevithick's vertical tubular boiler, 1830

non-conductor; the fourth annular space was to heat the air on its way to the fire; the fifth space for the used steam from the cylinder; lastly came the sixth space for again heating air to the fire. Distilled water was to be employed and there is provision for "make-up" water. Altogether he describes an admirable arrangement of boiler, condenser and heat interchanger. The ingenuity of this boiler and of the numerous others that Trevithick designed or constructed class him as the great pioneer of high-pressure boilers. There are very few of the thousand and one such boilers that have been designed or made since his time that do not embody ideas that he scattered with a lavish hand. In saying this, the authors are not claiming for Trevithick that he *originated* the tube boiler.

Friends exerted their influence on his behalf in his application to the Admiralty for the loan of a vessel and Giddy penned the following strongly worded testimonial, signed by a fellow Cornish M.P. and himself:[1]

We the undersigned having been requested to state our opinion of the Talents of Mr Richard Trevithick have no hesitation in recommending him as a man of extraordinary powers of mind and of fertility of invention.

Cornwall owes to him much of the Improvements that have been made on Mr Watt's engine, Improvements that have reduced the consumption of coals a third. Nor have his exertions been confined to steam engines alone.

He now proposes to make the same water act over and over again by alternate expansion and contraction which plan if it succeeds will be found of immense importance to steam vessels and to loco-motive engines.

Understanding that Mr Trevithick is desirous of making an engine on his new principle at his own risque, we strongly recommend him to the notice of the Lords of the Admiralty.

E. W. W. PENDARVES
DAVIES GILBERT.

Jan.y 27th 1832.

[1] Record Office, Admiralty Sec. In Letters, 4667, dated 1832, Jan. 27th, and cf. *Life*, II, 374.

Another friend who backed up Trevithick and found the money for his experiments was Mr George Mills. He wrote a letter,[1] signing it "For R. Trevithick", to the Admiralty enumerating six advantages of the engine and asking permission to build one for trial. Mills also wrote to the Royal Society two letters, one dated January 10th (a copy of the one above) and another dated March 2nd, in which he offers to "complete them [i.e. engines] at my own cost and risk not expecting any payment until the engines perform everything set forth in my agreement".

These letters were forwarded by the Royal Society to the Admiralty[2] and there is a copy of the reply to the Royal Society: "My Lords have offered Mr Mills a trial of his invention under certain conditions". Mr Mills likewise made friends with the technical officers, thus: "I am going to meet Capt. Symonds at Woolwich again tomorrow and hope to be able to persuade him to use his influence with Sir T. Hardy".[3]

On another occasion Mills writes: "I have just left Capt. Johnstone; he has communicated with Faucett & Co., Barnes and Miller and with the firm of Maudslay. He has had his mind disturbed again by Maudslay about the greater quantity of water required to condense steam at higher temperatures".[4]

We assume this to mean "at the higher ranges of temperature", and if so Maudslay was wrong, but he was only typical of manufacturing engineers at that day. It was ignorance of the properties of steam that was such a stumbling block in Trevithick's path. Remarks that he made to Goodrich at the interview quoted above may be reproduced: "The deeply-rooted prejudices entertained by the public that strong steam was not advantageous are now nearly removed by the superior performance of the Cornish engines over those of Boulton and Watt tho' it has required 26 years to partially open the eyes of the

[1] *Loc. cit.* 5061, dated 1832, Jan. 10th from Lauderdale House, Highgate.
[2] *Loc. cit.* 4282.
[3] *Life*, II, 339. Afterwards Admiral Sir William Symonds, Surveyor to the Navy 1832–1847; and Admiral Sir Thomas Masterman Hardy, first Secretary in 1838.   [4] *Life*, II, 375.

English engineers". Speaking of the provision of the steam gauge, a safeguard that he had consistently advocated, but which had not been adopted, Trevithick said: "So obstinate is John Bull that nothing short of an act of parliament will compel him to keep his brains safe in his own skull". What a delight it must have been to have sat at Trevithick's feet and listened to his racy language!!

Trevithick, while thus actively promoting his patent of 1831 now took out another [No. 6308], on September 22nd, 1832. The main claims will be understood by reference to the drawing accompanying the specification. These claims are for a manifold tubular superheater—and it will be observed what a modern look it has: the cylinder placed in the flue to keep it as hot as the steam; jet propulsion of vessels; application of the closed cycle boiler to a locomotive.

Francis Trevithick states that[1] in anticipation of the loan of a vessel from the Admiralty and to test the patents of 1831 and 1832, an engine of 100 horse-power was ordered in Shropshire; further, that a company known as the New Improved Patent Steam Navigation Company was formed in 1831, in which Trevithick was a shareholder. "This Company, among other proposals, opened negotiations for sending steam-boats to Buenos Ayres to help in the commerce of the port and inland river." To the very end Trevithick was busy on the idea of jet propulsion of vessels. The last letter of his we possess—to Giddy, May 14th, 1832—is full of it. The idea has been a dream of many inventors, both before and since, and it did not receive its quietus till the Admiralty trials of H.M.S. "Waterwitch" in 1868 showed that such a method was inferior in efficiency to others in use.

It is passing strange that nothing came, as far as we can learn, of all these efforts and of the provisional promise of the Admiralty to lend a vessel to try the engine. Trevithick had two purposes in view: one was to obviate loss of the latent heat necessitated by condensation; the other was to conserve feed water, so important an economy for steamships at a time when

[1] *Life*, II, 382.

boiler water was drawn from the sea and the engineer had to blow down about one-fifth of his boiler water to minimize incrustation. In doing this Trevithick re-invented the surface condenser, which, however, did not come into practical use for another decade. Nothing is more tantalizing in Trevithick's career than to find him on the threshold of big inventions and yet to know that he was just a few years too early to get them adopted. It has to be confessed that none of his efforts subsequent to his return from South America met with any success.

We have run on ahead somewhat, for we ought to have mentioned that Trevithick turned his attention in 1830 to warming houses by a portable hot-water stove. How the idea occurred to

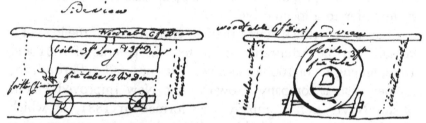

Fig. 36. Trevithick's apparatus for heating apartments, 1830

him is related in the following letter to Giddy, for whom Trevithick had been making a hot-house apparatus:[1]

While making a sketch of your work for the founder, a thought struck me that rooms might be better heated by hot water than by either steam or fire, and below I send you my thoughts on it, with a sketch, for your consideration. I fiend that steam pipes thats applied to heat cotton factorys, do heat with one surface foot of steam pipe, heat 200 Cubic feet of space to 60 degrees and keep up this temptre. I also found in Germany, where all the rooms are heated by the external sides of cast-iron pipes aboute the heat of steam, that one foot of external flue heated 160 Cubic feet of space to 70 degrs. I fiend also that aboute 200 surface feet of steam engine cylinder case supplied with steam will condence aboute as much steam as will produce aboute 15 Gallns of water per hour... and will consume aboute four bushells in 24 hours to

[1] *Enys Papers. Life*, II, 363. Letter dated Lauderdale House, 1830, March 1st.

keep it up to the tempture of 212 degrs. One bushell of coals will rise the tempture of 3600 pounds of water from 40 to 212 degrees.

A boiler, the same as the drawing,[1] will contain 1200 pounds of water, and would consume one third of a bushell of coals to raise this water from 40 to 212 degr. This boiler would have aboute 40 Surface feet of hot sides giving oute its heat. The 12 Inch fire tube in the boiler would get up to tempture from cold to 212 in abt 40 minutes from its own fuel. By these proofs it appears that 50 feet of surface steam sides will require one bushell of coals every 24 hours to keep up the boiling heat; therefore this boiler, having 40 Surface feet, would be near the surface sides required to give oute the heat of one third of a bushell of coals in 12 howers.... Now Suppose that this chardge of heat should be wanted to be thrown off in more or less than 12 hours, in that case the circular curtain would ajust the heat and time wanted for extracting this chardge of heat that was put into it by one third of a bushell of coals. By the foregoing experiments this coals and hot surface sides would heat to 60 degrs for 12 hours a space of 6800 Cubic feet, equal to a room of 25 ft square and 11 feet high.

If this boiler was placed in a room where there was a chimney, it might heat its water there withoute being removed by having a small shifting wrought Iron chimney tube of 4 Ins Diamr & 2 or 3 feet long attach'd to the end of the boiler while it was getting up the steam, after which it might be removed, and the doors at both ends of the boiler close shut; and as the boiler contains and retains its heat for 12 hours, more or less, it might be run on its wheels to any fire place or chimney to get charged with heat, and then run into any room, where there was no chimney, or in to a bead room, offices, or publick buildings; as it would be perfectualy free from risk, not having either steam or fire aboute it. The circular curtain, being fast to the wood table would, by being drawn up or down, ajust the required heat and hide the appearance of the boiler, and would be warm and comfortable to sit at. I think that this plan would save three quarters of the coal at present consum'd; the expence of the boiler alone would not exceed £5. 0. 0. When you have taken it into consideration, please to write me your opponion.

This is a valuable letter, and we have given it at length, as it reveals so well how Trevithick's mind worked and what good use he made of data that he, like every engineer, had accumulated. The inference from the words "I also found in Germany"

---

[1] The sketch is reproduced in Fig. 36.

is that he had been there, but if so we cannot say when or under what circumstances.

Doubtless Giddy's opinion, expressed in his reply, was that the idea was a good one and that he should go on with it. Giddy must have ordered one, for in a subsequent letter Trevithick writes:[1] "The last two or three days past I have been making experiments here on your warming machine from Cornwall, on which is engraved Davies Gilbert Esq. No. 1". The apparatus was $2\frac{1}{2}$ ft. diam. 3 ft. long and the fire tube 12 in. diam.; the results of two trials are reported. Trevithick pursued the matter and took out a patent [No. 6083] on the same day as the steam boiler patent, viz. February 17th, 1831. A drawing accompanying the patent specification delineates the apparatus substantially as described in the above letter to Giddy, and it will be readily understood without further remark.

In a later letter Trevithick says:[2]

I have already taken oute a patent in France for both the warming apparatus and engine. It have made a great bustle there amongest the French scientific class. When I see you, you shall know the particulears. I hope by this time you have fully tryd your warming machine, should be glad to know the result. There is a great many now in use in London and orders for more than I can execute. There is one at the George and Vulture in Cornhill of a Sirgosick [? sarcophagus] shape very handsomly brass ornimented. Its aboute $\frac{2}{3}$ the size of yours; it burns seven pounds of coals per day and keeps up the room to 65 degr. fifteen hours every day. The[y] are all nearly doing the same duty. The rage ammongst the ladys is to have them handsomly ornimented.

Coals at Paris is three shillings per hundred, four times the price that the[y] are in Cornwall. The Warming machines will sell there to a great extent. Som of them that was made in London goes for Paris tomorrow.

The letter hints that he was going in person to Paris on this errand. There is no trace of any French stove patent about this date, however. One would imagine that the stove would have found a better market on the Continent, where the closed stove

---

[1] *Enys Papers*. Letter dated Highgate, Lauderdale House, 1830, June 15th.
[2] *Enys Papers*. Letter to Giddy, dated Highgate, Lauderdale House, 1830, Dec. 31st.

is in general use, rather than in this country, where we adhere to yet shiver round the open fire. Although the stove was put upon the market, we have been unable to discover what if anything Trevithick made out of the invention.

In 1830 the rascality and extortion of the middlemen in the coal trade had attained such dimensions that the Corporation of the City of London took up the matter and appointed a committee on May 23rd, 1828, on "Charges upon Coals", which reported on September 3rd, 1830. Trevithick appeared before the Committee on February 27th, 1830, as he himself says:[1] and the minute of the Committee is as follows:[2] "Mr Trevithick an Engineer attending was heard in relation to having made a Steam Engine for weighing Coals of a portable nature which could be moved from Ship to Ship and that the same would cost about £105".

Although badly expressed, the meaning is obvious; we have no knowledge of what the apparatus consisted. The House of Commons, later, took up the same question of coal charges and it, too, appointed a Committee of the House, before which Trevithick appeared on May 4th. He sent in a bill to the Corporation for 5 guineas for each of these attendances, but the claim was disallowed on the ground that he attended at his own request.

Trevithick appeared before the Select Committee on Steam Carriages on August 12th and 17th, 1831, being in fact the fourth witness to be examined. He said: "I have noticed the Steam Carriages very much: I have been abroad for a good many years and had nothing to do with them till lately: but I have it in contemplation to do a great deal on common roads". He goes on to mention his recently patented engine as applicable to such carriages. He refers to his work on the Cornish engine; he mentions the Penydaren locomotive but not the London one and gives much more information, but says never a word about the Camborne and the London steam carriages, indeed his

[1] *Life*, II, 365.
[2] MS. City of London Record Office.

evidence, "I have had nothing to do with them [i.e. steam carriages] till lately", conveys to any ordinary reader of the evidence, and of Chapter II of this volume, either that he had forgotten about his trials in 1801 and 1803, or that he considered them too inconclusive to be worth mentioning. This omission is all the more extraordinary because Giddy was one of the Committee and could easily have put a leading question to him on the subject; it, therefore, looks as if Giddy too had forgotten Trevithick's early work.

Trevithick never troubled his head about politics, but that is not to say that he was not alive to the problems that at all times underlie the surface in progressive communities. Consequently the Reform Bills that were introduced in Parliament at this time aroused his sympathy, and when the first Reform Bill was introduced in March, 1831, he began upon the design of a colossal Reform Column of cast-iron. His original drawing, scale 50 feet to the inch, with a long explanation signed and dated "Lauderdale House, Highgate, May 5th, 1831", is preserved in his house at Penponds. The bill was thrown out by the Lords and led to a dissolution; the ensuing election resulted in an even more determined House of Commons, the throwing out of whose second bill led to such rioting that a third bill, carried by an enormous majority, after most dogged obstruction, received the Royal Assent on June 4th, 1832. In July a notice appeared in the newspapers of a public meeting to further the proposal of the erection of the enormous column. A number of noblemen and gentlemen signified their approbation and subscriptions limited to two guineas were invited.

A large lithograph, showing a plan and sectional elevation, was published. We cannot do better than give Trevithick's succinct description of the column:[1]

Design and specification for erecting a gilded conical cast-iron monument. Scale, 40 feet to the inch of 1000 feet in height, 100 feet diameter at the base, and 12 feet diameter at the top; 2 inches thick, in 1500 pieces of 10 feet square, with an opening in the centre of each

[1] *Life*, II, 390.

piece 6 feet diameter, also in each corner of 18 inches diameter, for the
purpose of lessening the resistance of the wind, and lightening the
structure; with flanges on every edge on their inside to screw them
together; seated on a circular stone foundation of 6 feet wide, with an

Fig. 37. Column to commemorate the passing of the Reform Bill,
1832, proposed by Trevithick. From a lithograph

ornamental base column of 60 feet high; and a capital with 50 feet
diameter platform, and figure on the top of 40 feet high; with a
cylinder of 10 feet diameter in the centre of the cone, the whole height,
for the accommodation of persons ascending to the top. Each cast-iron
square would weigh about 3 tons, to be all screwed together, with
sheet lead between every joint. The whole weight would be about

6000 tons. The proportions of this cone to its height would be about the same as the general shape of spires in England.

A steam-engine of 20-horse power is sufficient for lifting one square of iron to the top in ten minutes, and as any number of men might work at the same time, screwing them together, one square could easily be fixed every hour; 1500 squares requiring less than six months for the completion of the cone. A proposal has been made by iron founders to deliver these castings on the spot at 7*l*. a ton; at this rate the whole expense of completing this national monument would not exceed 80,000*l*.

There was to be an air lift to raise persons to the top.

A number of meetings were held during the next nine months and the plans were so far matured that Trevithick submitted the design to the King and received an acknowledgment on March 1st, 1833. But the shadow of death was already over Trevithick, and when the active brain was still, the movement stopped. To be quite candid, a column like this, more than five times that of the Nelson Column (184 ft. 10 in.), about four times the height of the Monument of the Great Fire of London, and more than two and a quarter times that of St Paul's, while interesting from the constructional side, as has been described, only panders to the gigantic in human psychology. Knowing what white elephants, or worse, these colossi become, we can only express satisfaction that the monument was not erected. The idea of erecting a National Column had, however, only been a side issue.

In connection with his experimental work, Trevithick had got into touch with John Hall, an enterprising young engineer who had recently taken over a millwright's business at Dartford, Kent, whither Trevithick now went. The tradition at the works, for it exists there to-day, is that Trevithick was designing and making a "turbine" engine. We suppose that it was a variant of his "recoil" engine or reaction turbine of 1815, and there is a drawing in existence, without date or description, which may represent this engine. He is said to have been experimenting also with steam boilers of 150 lb. pressure, but unfortunately no records of what he was doing have been preserved.[1] That he was working on the 1832 patent seems to be proved by the fact that Mr Edward Hall took out a patent in France for the same engine,[2] i.e. a "brevet d'importation", for fifteen years dated May 17th, 1833, for "une machine à vapeur agissant avec la vapeur dilatée". The drawing attached to the specification is identical with the drawing in the English one, but the patent, as may be inferred from the title, is restricted to the steam engine and boiler, and their application to marine propulsion. A reasonable inference is that an arrangement subsisted between Hall and Trevithick whereby the former should be accorded foreign, and perhaps, English patent rights, while the latter should have the run of the works for experiment.

But, alas, experiments and all mundane affairs were cut short by death. Trevithick died on the morning of April 22nd, 1833, after only a week's confinement to bed, at the Bull Inn, Dartford, where he was living. Mr Rowley Potter, a tradesman in the town, communicated the sad news to his family in Cornwall. His fellow workmen undertook the arrangements for his funeral, carried him to the grave on April 26th, and are said to have defrayed the funeral expenses, but probably John Hall did this. At Dartford the burial ground is not at the low-lying Parish Church but is some distance away, and is on an eminence where formerly stood a chapel dedicated to St Edmund, King and Martyr, known as "the upper church yard". Here, in a part reserved for the poor, he was buried; the exact spot is not known and no stone marks the place.[3] We are told that, to anticipate the

---

[1] Dunkin, *History and Antiquities of Dartford*, 1844, p. 406, says: "About 1832 having proposed the propulsion of vessels on rivers and canals by steam ejection of water in lieu of paddlewheels, the late Mr Hall invited him to Dartford to try the experiment on a vessel lately built; but after repeated trials, in which it is said upwards of £1200 was expended, the scheme was abandoned from the impossibility of obtaining pipes of sufficient strength to resist explosion". This statement is practically contemporary and therefore valuable but the concluding sentence reads like nonsense.

[2] *Brevets d'Inventions*, Ser. I, Vol. LXVII, p. 428, No. 9329.

[3] Dunkin, *loc. cit.*, p. 406, says that "the body of Mr Trevithick was deposited on the south side [of] the grave of Henry Pilcher. A subscription

nefarious practice of body-snatching then so prevalent, the coffin was provided with stout oak cross bearers above and below, connected by bolts with the nuts *underneath*, and, further, that night watchers remained by the grave.

The sole obituary notice we have been able to find is the following:

MR TREVITHICK. We regret to learn that this distinguished engineer, who may justly be regarded as the father of steam locomotion in England, died on the 22nd inst. at Dartford, in Kent, after a few days illness. He was in his 67th year.[1]

The cause of death was probably pneumonia, as frequently happens in the case of those who have returned from hot climates, accelerated by carelessness so common among men

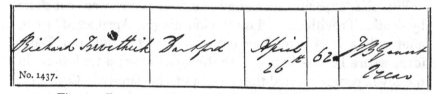

No. 1437.

Fig. 38. Register of Burials, 1833, Dartford Parish Church

who enjoy good health. The iron constitution that had stood him in such good stead had suffered serious inroads owing to his labours and privations in South America, but we learn nothing of any serious illness. It would seem that the reason why he made his headquarters at Highgate with Mr Gittins when engaged in London was the salubrity of the air there, for in a letter to Giddy he says:[2]

I have to appolige for my neglect for not calling on you [i.e. at Tredrea], but my ill health prevented it. I left home on the 11 of Feby, and arrived in town the 14th, and remained there untill the 24th, at which time I was compel'd to live for this place, this being a free good air. I am now taking twice a day the flours of Zinc, from

was commenced to erect a cast-iron monument over his remains, sufficiently elevated to be seen by passers by, but the funds raised were inadequate".
Mr Everard Hesketh in 1898 tried to locate the spot but without success.

[1] *Mech. Mag.* XIX, 80.

[2] *Enys Papers. Life*, II, 363. Letter dated Lauderdale House, Highgate, 1830, March 1st.

which I hope to be soon brought aboute again. I am much better but am as yet afraid to enter the City.

On April 12th he tells Giddy: "my health is restored".

He mentions attending a meeting at the Guildhall on February 27th but "with difficulty from my ill-health". About the time that he was attending the Committee, he wrote to his friend Gerard—and it must have been one of his last letters, for Gerard died shortly afterwards—about his complaint, which to judge by the description might have been asthma :[1]

Yesterday I took the coach to Highgate, by way of Camden Town and of course had to walk up Highgate Hill. I found I was able to walk up that hill with as much ease and speed as any of my coach companions. However strange this maggot may appear in my chest and brain it is no more than true. I wish among all you long-life-preserving doctors you could find out the cause of this defect, so as to remedy this troublesome companion of mine.

CHARACTER OF TREVITHICK

It is no easy task to analyse the character of such an erratic genius as was Richard Trevithick, full as it is of contrary elements defying our attempts to trace them to their origins. A man of simple tastes, genial, and with no trace of meanness in his nature, open-handed and open-hearted, his personality attracted people of all classes with whom he came in contact. Of splendid physique and tremendous nervous energy his confident self-reliance called for no man's support. He pursued his schemes with an overwhelming optimism, subject to rare periods of depression, which were due, however, to no want of faith in his own powers. No doubt he drew other men into disastrous speculations, but this was due to this same overpowering enthusiasm that enabled him to see nothing but the object of the moment. Of his various projects some were almost without foundation; others were based on sound reasoning which went directly to the core of the problem with the rapidity of sheer intuition. Rapid as this mental process was, it was acted upon almost as quickly. While other men deliberated Trevithick

----

[1] *Life*, ii, 388.

acted. Once having conceived an idea he put it into execution with a speed that astonishes us. With hardly any preliminary experiment he applied his engines in actual work to purposes hitherto untried. His indomitable forcefulness made him contemptuously intolerant of brains that moved more slowly than did his own. Yet though he quarrelled with others, he was a good fighter and a generous forgiver. His ire lasted only until the next wave of enthusiasm drowned it.

It is almost inevitable to contrast the mentalities of Watt and Trevithick. This is not that the one achieved worldly success when the other failed to do so, but while Watt was the great exponent of low pressure, Trevithick was the daring advocate of advance into a practically untried area of development. At the time they were both actively engaged in the same field, the one was a man whose life's work was nearing completion, and the conservatism of age deprecated the inevitable advance of the newer generation in ideas of which the other was the most advanced exponent.

Watt, one of our greatest men of science, was essentially the inductive philosopher. By nature retiring and unfitted to battle through life with his fellow-men, he had the nature of a recluse. His beautiful inventions were produced through a careful process of examination, of selecting the best of many ideas for effecting the desired object, and of self-doubting rejection of many of them by a painful process of elimination.

How different was the genius of Trevithick! With little theoretical knowledge, which indeed hardly existed in his day, his projects took practical form almost upon their conception. As a mechanical designer, if we allow for the state of the art of construction of his time, he stands pre-eminent; the very simplicity of his designs obscures the brilliancy of their conception. Throughout Trevithick's active life Giddy was his mentor, but only on matters of theory, in which he felt the lack of his own knowledge. On practical points he entertained no doubts, no self-worrying discussion of alternative methods of carrying an idea into effect, for there he felt sure of his own powers. Almost as intimate a correspondent on professional matters was

Rastrick; but here Trevithick always appears as the engineer directing the operative, while at the same time showing his confidence in Rastrick by leaving him much latitude. In the correspondences between these two men and himself we find much light thrown on the character of all three of them. With these exceptions Trevithick stands a very solitary and self-contained figure. In spite of the occasional bursts of confidential enthusiasm that show through his other correspondence, he pursues his way, ever scheming, ever feeling that some day he would achieve the elusive Eldorado that led him on in spite of his seeming disregard for money.

Watt brought his engine to a great perfection, but the low steam pressure at which it worked arrested its development; it remains one of the greatest of human achievements, a monument to his genius. To-day we build upon the foundations laid by Newcomen and Watt. But in Trevithick's ideas there was no finality—they were limited only by the materials and the state of the mechanic arts of his time. He was the prophet of high-pressure steam, the man of vision whom we are following to-day with our boilers and turbines, which generate and use steam of a pressure of hundreds of pounds to the square inch, while we look forward to still greater progress. And this is how Trevithick would have had it. He was intensely practical, yet a visionary; contemptuous of worldly goods, yet having a hope of a material reward. In the eager pursuit of his schemes and inventions—and it is here that we find the greatest difficulty in comprehending the man—all else—wife, family and friends— receded into the background. Thus we see him in Peru, after he had made money by the salving of the cargo of the "San Martin", when strongly advised to send home a thousand pounds or so to his wife, embarking it instead in a pearl fishery scheme, and losing it. Yet here appears the visionary. In spite of the rebuffs with which he had already met, was he not still in the land of the Incas—the land of fabulous wealth? To his inextinguishable optimism what was a paltry thousand or two to what he was going to do? Eldorado so often lies ahead.

Yet through it all—a great figure—a great Englishman.

Head and shoulders above his generation, in his strength and in his weaknesses, Trevithick stands a Colossus indeed. The neglect that he felt so keenly in life—are we not but recording it in the memorials we have at last erected or are erecting to his memory? Let us look upon them with humility, for they must ever remind us that in a nameless grave in a Kentish churchyard there lies a "heart once pregnant with celestial fire".

TREVITHICK PORTRAITURE

THE only contemporary portrait of Trevithick that we possess is the half length in oils painted by John Linnell (1792–1882) in London on the eve of Trevithick's departure for Peru. He was then forty-five years of age—in the prime of life—and is represented with the Andes in the background, to which he points as indicating his new sphere of activities. The portrait is arresting and vigorous and was considered by those who knew him to be a good likeness, and this is the impression that it gives; hence it has been chosen for the present volume (see Plate XV).

The portrait was in the possession of his widow till 1858, when Bennet Woodcroft, Clerk to the Commissioners of Patents, induced her to allow it to be included in the Gallery of Portraits of Inventors at the then Patent Office Museum, giving her in return a copy in oils and a photograph of the original. It is now one of the treasured portraits of the Science Museum, South Kensington, the successor to the Patent Office Museum.

The portrait of Trevithick in the group *Distinguished Men of Science living in 1808*, by William Walker, 1862, is from this painting.

Francis Trevithick states[1] that after his father's death casts were taken of the face and head, and that these were used for the marble bust executed prior to 1855 by Neville Northey Burnard, the Cornish sculptor. This bust is in the possession of Mr H. H. Trevithick, the present head of the family, and plaster copies are in existence (notably in the Royal Institution of Cornwall, Truro, presented in 1870 by W. J. Henwood, F.G.S. and in the Science Museum, South Kensington, presented in 1926 by Miss Trevithick). The face is clear cut, the proportions are noble and the whole gives the impression as being that of a man of great force of character. The differences between the bust

---

[1] *Life*, II, 245.

and the portrait are not greater than we should expect the ravages of time to have made in the interval between them.

There was a miniature on ivory of Trevithick as a young man, believed to be contemporary, in existence in 1858 and copies of it have been made. The original or a copy is in the possession of Captain R. E. Trevithick (see Plate II).

A bust in marble was commissioned in 1897 from C. H. Mabey, the sculptor, by the Institution of Civil Engineers, in whose rooms in London it is preserved. It is based on the portrait and bust already mentioned.

It had long been felt to be a reflection on the ingratitude we show to inventors generally and to Trevithick in particular that no statue of him should exist, particularly in his native county. This omission has now been remedied. In 1911 a few admirers in Cornwall formed a Trevithick Memorial Committee, for the purpose of erecting a statue and founding a scholarship. Subscriptions amounting to £600 had been received by 1914, but at the outbreak of the Great War efforts were relaxed. However, in 1917, Mr L. S. Merrifield, the sculptor, was commissioned to prepare a design and completed a model which he exhibited in the Royal Academy in the following year. Through the efforts of Mr J. M. Holman, and other members of the Committee, the necessary amount, enhanced by the rise in prices generally, was obtained, and the statue was unveiled at Camborne, by H.R.H. Prince George, on May 17th, 1932, in the presence of a large crowd. The statue stands at the cross roads outside the Passmore Edwards Free Library and faces Beacon Hill, the scene of the road locomotive trials of 1801. The statue is in bronze, larger than life size, i.e. 7 ft. 6 in. high, and stands on a plinth of Cornish granite with bronze panels on all four sides, those on the front and sides depicting his inventions, and on the back, passages in his life. The sculptor has succeeded not only in giving us a good portrait, but in the pose of the figure has conveyed the impression of the energy and virility of the man. Trevithick holds in his left hand, very appropriately, a model of his road locomotive, to which he is pointing with his right, while gazing up Beacon Hill. On the base is the name

PLATE XVII. RICHARD TREVITHICK
STATUE AT CAMBORNE, UNVEILED 1932

Photo by Bennetts, Camborne

"Trevithick" and, at the back, "Richard Trevithick, inventor of the locomotive engine. Born 1771. Died 1833". The panels represent allegorical female figures holding: (1) front, model of the Cornish boiler; (2) left side, model of the South Wales rail locomotive, and a modern engine in the background; (3) right side, model of the ladder dredger and a modern liner in the background; (4) rear, a sailing ship with the sun rising behind it; on the right hand is Carn Brea and on the left the mule track over the Andes. We have given a description of this statue at some length because, in its actuality and by its symbolism, it conveys so well to the onlooker the story of Trevithick—a consummate artistic achievement.

Prior to the war the same sculptor was commissioned by the late Lord Rhondda to execute a statue for Merthyr Tydvil to mark the rail locomotive trials there in 1804, but owing to his lordship's death the project was abandoned (but cf. p. 266).

MEDALLIC PORTRAITS

The Grand Trunk Railway of Canada instituted, in 1860, an annual efficiency and good conduct medal in bronze for their servants.

> Head of Trevithick *l.*; at *r.* side "Richard Trevithick", below "J. S. Wyon, Sc."; around "Grand Trunk Railway of Canada". *Rev.* "Presented by the Directors to
> for general efficiency and good conduct during the year ".
> 45 mm. diam. Silver, by Joseph Shepherd Wyon (1836–73).

MEMORIALS TO RICHARD TREVITHICK

IN view of the neglect of Trevithick during his lifetime and the oblivion into which he seemed to pass on his death, it occasions no surprise to learn that for more than half a century after that event no memorial of him existed anywhere. Perhaps there may have existed some sentiment of the proud kind voiced by his son Francis in 1858, who, writing[1] of his father's burial-place, said: "The grave is one of those given to the unknown, no tablet speaks for the dead, a pauper's turf is the only mark of Richard Trevithick and to my mind more in keeping with his singleminded original genius than the gilded slab". The authors prefer to look upon this lack of memorials rather as one simply of neglect and ignorance.

Tardy recognition came in 1883, when Hyde Clarke, among other activities Editor of the *Railway Review*, a great admirer and the earliest biographer of Trevithick, suggested in the *Mining Journal* (1882, Oct. 22nd, p. 1275) that to mark the fiftieth anniversary of Trevithick's death, a memorial, to take the form of a statue in Westminster Abbey, should be erected. A meeting was held on April 10th, 1883, at the Institution of Civil Engineers, when it was "Resolved to raise a Fund for the erection of a statue to the memory of Richard Trevithick and further to provide a Fund for the Establishment of Scholarships bearing his name, such Scholarships to aid in the Technical Education of Young Men to qualify them for the profession of Mining or other Engineers", and also that "Employers of labour be requested to allow Penny Subscriptions from their workmen". A large and influential Committee, with a branch in Cornwall, was formed; H.R.H. the Prince of Wales was at its head and Major (afterwards Lt.-Col.) John Davis, F.S.A., was Hon. Sec. *A Memorial edition of the life of Richard Trevithick—* really an abridgement of the large work of Francis Trevithick—

---

[1] Holograph. L. F. Trevithick to Bennet Woodcroft, 1858, June 15th, in the Science Museum, South Kensington.

was published to help the movement. Subscriptions, even the pence of the workmen in engineering shops, were invited, but money came in slowly and it became evident that the sum would be insufficient for the whole of the proposed scheme. In consultation with the Dean and Chapter of Westminster, where it had been hoped to erect the statue, and following precedent in the case of Stephenson, Brunel and other engineers, a memorial window was decided upon. A window opening in the north aisle was allocated and the stained glass placed in position in the spring of 1888. A meeting of subscribers to view the finished memorial was held on June 13th.

The window is a two lancet one with quatrefoil tracery above. In the lancets, nine figures of Cornish saints are represented in two tiers of canopied niches. In the upper tier are St Piran, St Petroc, St Pinnock and St Germains, with their names underneath. In the lower tier are St Julian with St Cyriacus, St Constantin, St Nonna and St Geraint, similarly named. Below these again is a third tier of niches with figures of angels holding scrolls on which are outline drawings of Trevithick's inventions: viz. "Tramroad Locomotive 1803", "Cornish Pumping Engine", "Steam Dredger 1803", "Railway Locomotive 1808". In canopies at the head of the lights are shields blazoning the arms of the see and duchy of Cornwall respectively. At the foot is the name of the inventor with dates of his birth and of his death.

The window is rather dark and although, perhaps, no better symbolism was to be expected at this period, it strikes the beholder as hardly worthy of the great engineer. It is a pity too that so few people see it. The cost of the window was £1066 and with further sums collected a Trevithick Scholarship was endowed at a cost of £1000 at Owens College, Manchester, to be awarded triennially. The balance of the fund, a sum of £100. 6s., was placed in trust for the award of a Trevithick Premium by the Institution of Civil Engineers. The scholarship was first awarded in 1889 and the premium in the session 1900–1901,[1] and since then triennially and biennially respectively.

It is of interest to know that the Trevithick Premium fund was

[1] *Proc. Inst. C.E.* LII, 198.

augmented by £500 in 1932 by Dr H. K. Trevithick, J.P., in memory of her husband, grandson of the inventor.

A bronze mural tablet was placed on the garden wall of Mr Holman's house close to the Weith, Camborne, where the road locomotive was built, on the occasion of the hundredth anniversary of the trial, December 23rd, 1901, by a small band of admirers.

A bronze memorial tablet, designed by Mr W. S. Rogers to mark Trevithick's connection with Dartford, was set up in the south aisle of the Parish Church of Holy Trinity, by the generosity of Mr Everard Hesketh of the firm of Messrs J. and E. Hall (see Plate XVIII) and dedicated on March 9th, 1902, by the Rt. Rev. the Bishop of Dover.

We have already mentioned the statue at Camborne, and it remains to say that a scholarship at the School of Metalliferous Mining, Camborne, in affiliation with the Victoria University of Manchester, was part of the scheme. Largely with funds collected in South Africa by Mr J. M. Holman, this "Trevithick Memorial Scholarship" was established in 1923, and has been awarded annually since. It covers the provision of free instruction and £30 for maintenance per annum.

To decide what should be done to mark the Centenary of Trevithick's death, a meeting of members of the engineering profession was held at the Institution of Civil Engineers on October 20th, 1932, when it was resolved: "That this meeting of members of the engineering profession decides to honour in 1933 the memory of Richard Trevithick on the occasion of the centenary of his death".

The name "Trevithick Centenary Commemoration" was adopted as the title of the body set up at the meeting; it was decided that its scope should be international, and that foreign engineers and governments should be invited to take part. A number of proposals were made, e.g. the erection of memorials at his birthplace, in London and at Merthyr Tydvil, a Centenary Exhibition of Trevithick objects, Memorial Services and Lectures and the publication of a Memorial Volume. An Executive Committee was appointed to carry out these proposals. The

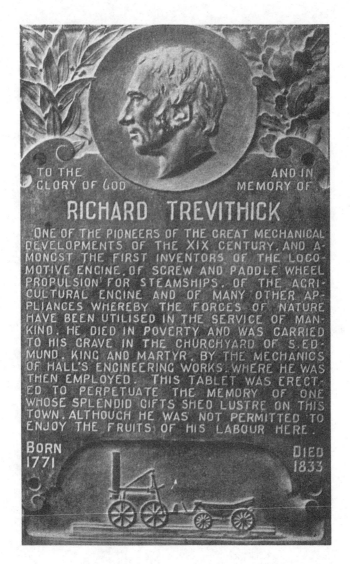

PLATE XVIII

MEMORIAL TABLET TO RICHARD TREVITHICK
IN DARTFORD PARISH CHURCH, 1902

Courtesy of Everard Hesketh, Esq.

Exhibition was held at the Science Museum, South Kensington, from March to June, and quite a remarkable number of exhibits was got together. The Memorial Services were held at Dartford Parish Church and at Westminster Abbey on April 23rd. The Memorial Lecture was delivered by Prof. C. E. Inglis, F.R.S. on April 24th, at the Institution of Civil Engineers. The Memorial Volume is, as explained in the Preface, the present one. While it is passing through the press, we learn that the memorials in London and Merthyr Tydvil are rapidly nearing completion.

LETTERS PATENT FOR INVENTIONS

APPENDIX I

TREVITHICK AND VIVIAN'S PATENT
SPECIFICATION, 1802

STEAM ENGINES ⎫
Improvements in the ⎪
construction thereof ⎪
and Application there- ⎬
of for driving ⎪
Carriages ⎪
Trevithick & Vivian's ⎪
Specification ⎭

TO ALL WHOM these presents shall come I ANDREW VIVIAN of the parish of Camborne in the County of Cornwall Engineer and Miner Send Greeting. WHEREAS his present most excellent Majesty King George the Third Did by his Royal Letters under the Great Seal of the United Kingdom of Great Britain and Ireland bearing Date at Westminster the twenty fourth day of March in the forty second year of his Reign Give and Grant unto Richard Trevithick of the parish of Camborne aforesaid Engineer and Miner and me the said Andrew Vivian our Executors Administrators and Assigns His especial Licence full power sole Privilege and Authority That we the said Richard Trevithick and Andrew Vivian our Executors Administrators and Assigns during the term of years therein expressed should and lawfully might make use exercise and vend our Inventions of Methods for improving the construction of Steam Engines and the application thereof for driving Carriages and for other purposes within England Wales and the Town of Berwick upon Tweed and the Colonies plantations and Dominions abroad in such manner as to us the said Richard Trevithick and Andrew Vivian our Executors Administrators and Assigns should in our Discretion seem meet. IN WHICH said Letters Patent is contained a proviso that if we the said Richard Trevithick and Andrew Vivian or one of us should not particularly describe and ascertain the nature of our said Invention and in what manner the same is to be performed by an Instrument in writing under our Hands and Seals or under the Hand and Seal of one of us and cause the same to be inrolled in His Majesty's High Court of Chancery within one calendar Month next and immediately after the Date of the said Letters Patent That then the said Letters Patent and all Liberties and Advantages whatsoever thereby granted should utterly cease determine and become void As in and by the said recited Letters

Patent (Relation being thereunto had) may most fully and at large appear. NOW KNOW YE that in compliance with the said proviso I the said Andrew Vivian do hereby declare that our said Invention is described in manner following that is to say:

"Our improvements in the construction and application of steam-engines are exhibited in the drawings hereunto annexed and explained, namely:—

"In Plate IV, Figure 1[1] represents the vertical section of a steam-engine with the said improvements; and Figure 2 represents another vertical section of the same engine at right angles to the plane of Figure 1. The dark shaded parts represent iron, and the red parts represent brickwork, and the yellow parts engine brass, excepting only the wooden supporters of the great frame in Figures 4 and 5, and the carriage-wheels in 6 and 7. A represents the boiler made of a round figure, to bear the expansive action of strong steam. The boiler is fixed in a case D, luted inside with fire-clay, the lower part of which constitutes the fire-place B, and the upper cavity affords a space round the boiler, in which the flame or heated vapour circulates round till it comes to the chimney E. The case D and the chimney are fixed upon a platform F, the case being supported upon four legs. C represents the cylinder enclosed for the most part in the boiler, having its nozzle, steam-pipe, and bottom cast all in one piece, in order to resist the strong steam, and with sockets in which the iron uprights of the external frame are firmly fixed. G represents a cock for conducting the steam, as may be more clearly seen by observing Figure 3, which is a plan of the top of the cylinder, and the same parts in Figure 2.

"b, Figures 2 and 3, represents the passage from the boiler to the cock G. This passage has a throttle-valve or shut, adjustable by the handle m, Figure 2, so as to wiredraw the steam, and suffer the supply to be quicker or slower. The position of the cock represented in Figure 3 by the yellow circle is such, that the communication from the boiler through b, by a channel in the cock, is made good to d, which denotes the upper space of the cylinder above the piston, at the same time that the steam-pipe a (more fully represented in Figure 1) is made to afford a passage from the lower space in the cylinder beneath the piston to the channel C, through which the steam may escape into the outer air, or be directed and applied to heating fluids or other useful purposes. It will be obvious that if the cock be turned one quarter of a

---

[1] Figures 39 to 41 have been dissected from the Patent Specification drawing (see p. 53) in the form here given by the kind offices of our friend Mr W. J. Tennant, C.P.A.

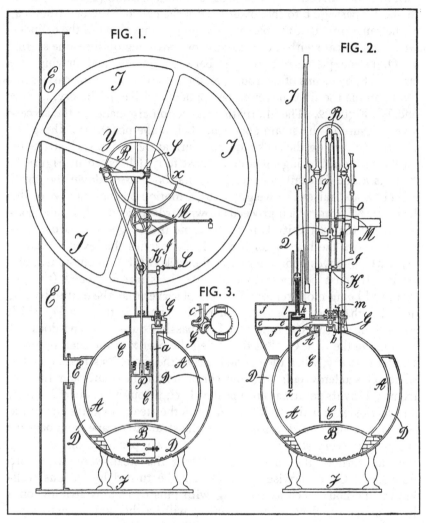

Fig. 39. Trevithick and Vivian's high-pressure steam engine. From their patent specification, 1802

turn in either direction, it will make a communication (Figure 3) from the boiler passage b to the lower part of the cylinder, by or through a, at the same time that the passage r from the upper part of the cylinder will communicate with c, the passage for conveying off the same steam. P, Q, is the piston-rod moving between guides, and driving the crank R, S, by means of the rod Q, R, the axis of which crank carries the fly T, and is the first mover to be applied to drive machinery, as at S and W, Figure 2. The alternations of action are made by the successive pressure of the steam above and below the piston, and these are effected by turning the cock a quarter turn at the end of each stroke by means of the following apparatus, most fully delineated in Figure 1. x, y, is a double snail, which in its rotation presses down the small wheel O, and raises the weight N by a motion on the joint M of the lever O, N, from which proceeds downwards an arm, M, L, and consequently the extremity L is at the same time urged outwards. This action draws the horizontal bar L, I, and carries the lever or handle H, I, which moves upon the axis of the cock G through one-fourth of a circle. It must be understood that H, I, is foreshortened (the extremity I being more remote from the observer than the extremity H), and also that there is a click and ratchet-wheel in the part H, which gathers up during the time that L is passing outwards, and does not then move the cock G. But that when the part x of the snail opposite O, that is to say, when the piston is about the top of its stroke, then the wheel O suddenly falls into the concavity of the snail, and the extremity L by its return at once pushes I, H, through the quarter circle, and carries with it the cock G, and turns the steam upon the top of the piston, and also affords a passage for the steam to escape from beneath the piston; every stroke, whether up or down, produces this effect by the half turn of the snail, and reverses the steam-ways as before described. Or, otherwise, the cock may be turned by various well-known methods, such as the plug with pins or clamps striking on a lever in the usual way, and the effect will be the same, whether the quarter turns be made back or forward, or by a direct circular motion, as is produced by the machinery here delineated, but the wear of the cock will be more uniform and regular if the turns be all made the same way. In the steam-engines constructed and applied according to our said Invention, the steam is usually let off or conducted out of the engine, and in this case no vacuum is formed in the engine, but the steam, after the operation, is or may be usefully applied as before mentioned. But whenever it is found convenient or necessary to condense the steam by injection water, we use a new method of condensing by an injection above the bucket of the air-pump; and by this Invention we render the condenser or space which is usually constituted or left between the said bucket and the foot-valve entirely unnecessary, and we perfectly exclude the admission of any elastic fluid from the injection water into the internal working spaces of the engine.

"In Figure 2 is represented a method of heating the water for feeding the boiler, by the admission of steam after its escape through c into the cistern f. The steam passes under a false bottom e, perforated with small holes, and heats the water therein, a portion of which water is driven at every revolution of the fly by the small pump k, through l, z, into the boiler A. We also on some occasions produce a more equable rotary motion in the several parts of the revolution of any axis moved by steam-engines, by causing the piston-rods of two cylinders to work on the said axis by cranks at one quarter turn asunder. By this means the strongest part of the action of one crank is made to assist the weakest or most unfavourable part of the action in the other, and it becomes unnecessary to load the work with a fly.

"Figure 4 is an upright section, and Figure 5 is a plan of the engine, with rollers for pressing or crushing sugar-canes, moved by a steam-engine improved and applied according to our said new Invention. B is a case, in the form of a drum or cylinder, suspended upon two strong trunnions or pivots at O and O, its flat ends standing upright; within the iron case is fixed a boiler A, not much smaller in its dimensions, but so as to leave a vacant space between itself and the case, and within the boiler is fixed a fire-place, having its grate above the ash-hole D; the heated vapour and smoke rises at the inner extremity, and passes through two flues E, E, Figure 5, which join above at E, m, Figure 4, in the chimney E, which is there loosely applied, and is slung between centres in a ring at F. The working cylinder C, with its piston, steam-pipe, nozzle, and cock, are inserted in the boiler as here delineated. The piston-rod drives the fly T, T, upon the arbor of which is fixed a small wheel, which drives a great wheel upon the axis of the middle roller. The guides are rendered unnecessary in this application of the steam-engine, because the piston-rod is capable, by a horizontal vibratory motion of the whole engine upon its pivots O, to adapt itself to all the required positions, and while the lower portion of the chimney E, m, Figure 4, partakes of this vibratory motion, the upper tube E, F, is enabled to follow it by its play upon the two centres or pivots in the ring F. In such cases or constructions as may render it more desirable to fix the boiler with its chimney and other apparatus, and to place the cylinder out of the boiler, the cylinder itself may be suspended for the same purpose upon trunnions or pivots in the same

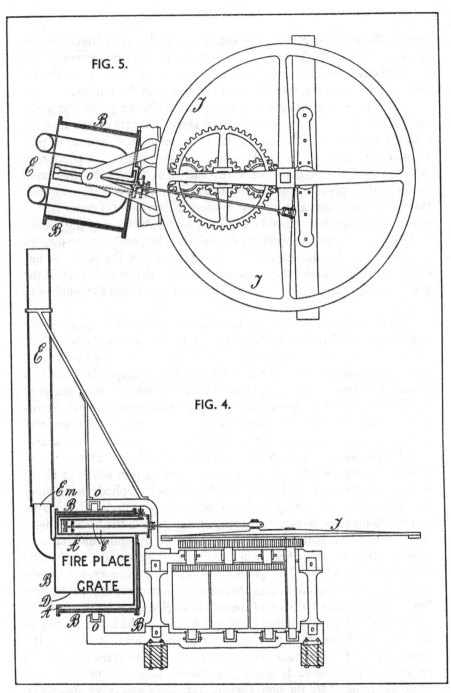

FIG. 5.

FIG. 4.

FIRE PLACE

GRATE

Fig. 40. Trevithick and Vivian's high-pressure steam-engine and
sugar mill. From their patent specification, 1802

manner, one or both of which trunnions or pivots may be perforated so as to admit the introduction and escape of the steam, or its condensation as before mentioned. And in such cases, when it may be found necessary or expedient to allow of no vibratory motion of the boiler or cylinder, the same may be fixed, and the method of guides be made use of, as in Figure 1 or 2. The manner in which the cock is turned is not represented in these two drawings, but every competent workman will, without difficulty, understand that this effect may be produced by the same means as in Figure 1, or otherwise by the stroke of pins duly placed in the circumference of the fly, and made to act upon a cross fixed on the axis of the cock, or otherwise by the method used in the carriage, Figure 6, and hereinafter described. The steam which escapes in this engine is made to circulate in the case round the boiler, where it prevents the external atmosphere from affecting the temperature of the included water, and affords, by its partial condensation, a supply for the boiler itself, and is or may be afterwards directed to useful purposes as aforesaid.

"Figure 6 is a vertical section, and Figure 7 the plan of the application of the improved steam-engine to give motion to wheel-carriages of every description. B represents the case, having therein the boiler with its fire-place and cylinder, as have been already described in Figure 4.

"The piston-rod P, Q, Figure 7, is divided or forked, so as to leave room for the motion of the extremity of the crank R. The said rod drives a cross-piece at Q backward and forward between guides, and this cross-piece, by means of the bar Q, R, gives motion to the crank with its fly F, and to two wheels T, T, upon the crank axis, which lock into two correspondent wheels U upon the naves of the large wheels of the carriage itself. The wheels T are fitted upon round sockets, and receive their motion from a striking box or bar S, X, which acts upon a pin in each wheel. S, Y, are two handles, by means of which either of the striking boxes S, X, can be thrown out of gear, and the correspondent wheel W by that means disconnected with the first mover, for the purpose of turning short or admitting a backward motion of that wheel when required. But either of the wheels W, in case of turning, can be allowed considerably to overrun the other without throwing S, X, out of gear, because the pin can go very nearly round in the forward motion before it will meet with any obstruction. The wheels U are most commonly fixed upon the naves of the carriage-wheels W, by which means a revolution of the axis itself becomes unnecessary, and the outer ends of the said axis may consequently be set to any obliquity, and the other part fixed or bended, as the objects of taste or utility may demand. The fore wheels are applied to direct the carriage by means of a lever H, and there is a check-lever which can be applied to the fly, in order to moderate the velocity of progression while going down hill. In the vertical section r, u, denotes a springing lever, having a tendency to fly forward. Two levers of this kind are duly and similarly placed near the middle of the carriage, and each of them is alternately thrown back by a short bearing lever S, t, upon the crank axis, which sends it home into the catch u, and afterwards disengages it when the bearing lever comes to press upon V, in which case the springing lever flies back. A cross-bar or double handle o, p, is fixed upon the upright axis of the cock, from each end of which said cross-bar proceeds a rod p, q, which is attached to a stud q, that forms part of the springing lever r, u. This stud has a certain length of play, by means of a long hole or groove in the bar, so that when the springing lever r, u, is pressed up, the stud slides in the groove without giving motion to p. When the other springing lever is disengaged, it draws the opposite end of p, o, by which means p draws the long hole at q, up to its bearing against the stud, ready for the letting off of that first-mentioned springing lever. When this last-mentioned lever comes to be disengaged, it suddenly draws p back, and turns the cock one quarter turn, and performs the like office of placing the horizontal rod of the other extremity of p, o, ready for action by its own springing lever. These alternations perform the opening and shutting of the cock, and to one of the spring levers, r, u, is fixed a small force-pump w, which draws hot water from the case by the quick back-stroke, and forces it into the boiler by the stronger and more gradual pressure of S, t. It is also to be noticed, that we do occasionally, or in certain cases, make the external periphery of the wheels W uneven, by projecting heads of nails or bolts, or cross-grooves, or fittings to railroads when required; and that in cases of hard pull we cause a lever, bolt, or claw, to project through the rim of one or both of the said wheels, so as to take hold of the ground; but that in general the ordinary structure or figure of the external surface of these wheels will be found to answer the intended purpose. And, moreover, we do observe and declare, that the power of the engine, with regard, to its convenient application to the carriage, may be varied, by changing the relative velocity of rotation of the wheels W, compared with that of the axis S, by shifting the gear or toothed wheels for others of different sizes properly adapted to each other in various ways, which will readily be adopted by any person of competent skill in machinery. The body of the carriage M may be made of any convenient size or figure, according to its intended uses.

"And lastly, we do occasionally use bellows to excite the fire, and the said bellows is worked by the piston-rod or crank, and may be fixed in any situation or part of the several engines herein described, as may be found most convenient."

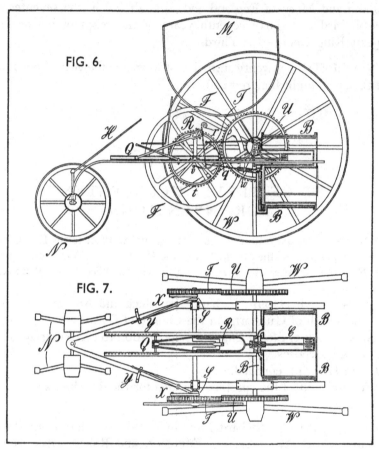

Fig. 41. Trevithick and Vivian's steam road carriage.
From their patent specification, 1802

IN WITNESS whereof the said Andrew Vivian have hereunto set my hand and seal this twenty sixth day of March in the year of our Lord One thousand eight hundred and two

ANDREW (L S) VIVIAN

AND BE IT REMEMBERED that on the twenty sixth day of March in the forty second year of the reign of his Majesty King George the Third the said Andrew Vivian came before our said Lord the King in his Chancery and acknowledged the Instrument aforesaid and all and everything therein contained and specified in form above written and also the Instrument aforesaid was stamped according to the tenor of the several statutes made in the sixth year of the late King and Queen William and Mary of England and so forth and in the seventeenth twenty third and thirty seventh years of the reign of his present Majesty King George the Third.

INROLLED[1] the twenty sixth day of March in the year of our Lord one thousand eight hundred and two.

APPENDIX II

LIST OF TREVITHICK'S PATENTS
FOR INVENTIONS

* No patent on the roll. Entry of sign manual warrant to the Law Officer for the preparation of the Patent Bill in the Home Office Warrant Book.
† No patent on the roll. No entry in the Home Office Warrant Book.

1802, 24 March [2599]   Richard Trevithick and Andrew Vivian of the parish of Camborne in the County of Cornwall.
Construction of Steam Engines. Application to drive carriages and other purposes.
Large Drawing enrolled.
(For full text of the specification, and for the drawing, see Appendix I and p. 53.)

1808, 5 July [3148]   Richard Trevithick of the parish of Rotherhithe, in the County of Surrey, Engineer, and Robert Dickinson of Great Queen Street, in the County of Middlesex, Esquire.
Machinery for towing, driving or forcing and discharging ships and other vessels of their cargo.
"The novelty of our invention is simply this: Employing such a vessel as we have described, furnished with a steam engine and moving power, and with proper apparatus to enable us to employ the said vessel and its contents as a labourer, to assist in towing of vessels in the manner before described and in loading and unload-

[1] The enrolled drawing is on parchment by "John Newton, Chancery Lane", and measures 2 ft. 8 in. by 2 ft. 3 in.

ing them in place of using the methods hitherto in use." When
only one function is to be performed the apparatus may be relieved
of the unwanted part.

No Drawing enrolled.

1808, 31 Oct. [3172]   Richard Trevithick and Robert Dickinson (addresses, etc., as in 3148).

Stowing ships' cargoes by means of packages to lessen expense of
stowage, and keep the goods safe. "We do construct make use
and apply certain other packages vessels or receptacles of iron",
i.e. iron tanks.

No Drawing enrolled.

1809, 29 April [3231]   Richard Trevithick and Robert Dickinson (addresses, etc., as in 3148).

Naval architecture and navigation applicable to other purposes.

The claims are (1) floating docks, (2) iron ships for ocean service,
(3) masts, etc., of iron, (4) bending timber, (5) diagonal framing
for ships, (6) iron buoys, (7) steam engines for general ships use,
(8) rowing trunk, (9) steam cooking.

No Drawing enrolled.

1810, 23 March [*]   Richard Trevithick of Fore Street, Limehouse,
in the County of Middlesex, Engineer, and Robert Dickinson, of
Great Queen Street, in the County of Middlesex, Esquire.

Inventions or new applications of known Powers to propel Ships &
other Vessels employed in Navigating the Seas or in Inland
Navigation to aid the recovery of Shipwrecks, promote the health
& comfort of the Mariners and other useful purposes.

1815, 6 June [3922]   Richard Trevithick of the parish of Cambrone
(*sic*) in the county of Cornwall, Engineer.

High Pressure Steam Engine; and application thereof with or without other machinery to useful purposes.

No Drawing enrolled.

1815, 20 Nov. [3922]   A second specification, same address, states:
"A part of the invention having been through error omitted in the
said specification [i.e. the first one], I am now desirous of
correcting the same".

The claims are (1) plunger pole steam engine, (2) reaction turbine
or aeolipile, (3) high pressure steam acting on water which acts
on a piston, (4) the water from (3) used in (2) as in a barker's
mill, (5) screw propeller.

1816, 22 June [*]   Richard Trevithick of Penzance in the County of Cornwall, Engineer.

A new apparatus for evaporating water from solutions of vegetable substances.

1827, 10 Nov. [†]   Richard Trevithick in the Parish of Saint Erth in the County of Cornwall, Civil Engineer.

New methods for centering ordnance on pivots, facilitating the charge of the same and reducing manual labour in time of action.

1828, 27 Sept. [*]   Richard Trevithick of Saint Erth in the County of Cornwall but at present residing at Saint Mary Axe in the City of London, Civil Engineer.

Certain new methods of machinery for discharging Ships' cargoes & other purposes.

1829, 27 March [*]   Richard Trevithick of the Parish of Saint Erth in the County of Cornwall, Civil Engineer.

A new or improved steam engine.

(Doubtless this was dropped in favour of No. 6082 below.)

1831, 21 Feb. [6082]   Richard Trevithick of the parish of Saint Aith (*sic*) in the County of Cornwall, Engineer.

Steam engine.

The claims are (1) boiler and condenser, (2) condenser in air vessel, (3) surface condenser, (4) condensed water returned to boiler, (5) forced draught with hot air heated by condenser water.

No Drawing enrolled.

1831, 21 Feb. [6083]   Richard Trevithick (address, etc., as in 6082).

Apparatus for heating apartments.

Consists of a portable stove surrounded by water brought to the boiling point.

Drawing enrolled.

1832, 22 Sept. [6308]   Richard Trevithick of the parish of Cambourne, in the County of Cornwall, Engineer.

Steam engine; application of steam power to navigation and locomotion.

The claims are (1) super-heater, (2) cylinder kept in flue to be hotter than steam, (3) jet propulsion of vessels, (4) boiler and super-heater applied to a locomotive.

One Sheet of Drawings enrolled.

BIBLIOGRAPHY

MANUSCRIPTS

1775 onwards   BOULTON AND WATT COLLECTION. The well-known collection in the Reference Library, Birmingham. Quoted by the authors as *B. and W. Colln.*

1780–1803   BOULTON AND WATT LETTERS. About 1500 letters mounted in twelve quarto volumes, being the correspondence of the firm with Thomas Wilson, their agent in Cornwall, now in the possession of the Royal Cornwall Polytechnic Society at Falmouth. Quoted by the authors as *B. and W. Letters.*

1797–1816   LETTER AND ACCOUNT BOOKS. Four parchment-covered mine account and cost books which had belonged to Richard Trevithick senior, the unused portions used by the son as draft letter books and account books; in the possession of Capt. R. E. Trevithick. Quoted by the authors as *Draft Letter Books* and *Rough Account Books.*

1799–1836   TREVITHICK-GIDDY CORRESPONDENCE. A foolscap volume containing sixty-four letters of Trevithick and other cognate matter, constituting the largest source of information about him. Preserved by the Enys family at Enys and now in the possession of the Royal Institution of Cornwall, Truro. Quoted by the authors as *Enys Papers.*

PRINTED MATTER

1809   NICHOLSON, WILLIAM. *British Encyclopedia*, Vol. III, 1809, Art. "Engine for raising water". Describes and illustrates Trevithick's water-pressure engine at Wheal Druid.

1809   TREVITHICK, RICHARD, and DICKINSON, ROBERT. *Outlines of a Patent obtained for an improvement in the Stowage of Ships for preserving their cargoes, naval and military stores, etc., etc., etc.* London, 1809, 12mo. At the same date the patentees also published a similar pamphlet describing their floating dock. The authors have not seen a copy.

1817   HOUSE OF COMMONS. *Select Committee on Steam boats.* London, 1817, fol. Evidences the opinions held at the time on boilers and boiler materials.

1818   BOAZE, HENRY. "On the introduction of the Steam engine to the Peruvian Mines." *Trans. Royal Geol. Soc. of Cornwall*, Vol. I, p. 212. Camborne, 1818, 8vo. Based on first-hand information obtained from the Trevithick family.

1819   REES's CYCLOPAEDIA. Art. "Steam Engine", Plate IX, and Art. "Hydraulics", Plate III. London, 1819, 4to. In the first article Trevithick's engine is described; in the second his dredger is described but without mentioning his name.

1824 STUART, ROBERT (pseudonym of MEIKLEHAM, ROBERT). *Descriptive History of the Steam Engine.* London, 1824, 8vo. Trevithick's engine is described, pp. 162, 182.

1825 STEVENSON, WILLIAM BENNET. *Historical and Descriptive Narrative of twenty years residence in South America.* London, 1825, 3 vols. 8vo. The author was in Chile and Peru during the War of Secession and held official positions there. He was private secretary to Lord Cochrane.

1826 MIERS, JOHN. *Travels in Chile and La Plata.* London, 1826, 8vo. The author was personally acquainted with the chief actors in Trevithick's Peruvian venture.

1827 GILBERT, DAVIES, P.R.S. "Observations on the Steam Engine." *Phil. Trans.* Jan. 25th, 1827. London, 1827, 4to. A review of the position.

1829 STUART, ROBERT (pseudonym of MEIKLEHAM, ROBERT). *Historical and Descriptive Anecdotes of Steam Engines and of their Inventors and Improvers.* London, 1829, 2 vols. 16mo. Trevithick's engines are described in Vol. II, pp. 458–460, and 515–520.

1831 GALLOWAY, ELIJAH. *History and Progress of the Steam Engine...to which is added an extensive appendix...by Luke Hebert.* London, 1831, 8vo.

1832 TREVITHICK, RICHARD. Design for a gilded national monument of cast-iron 1000 feet high and 100 feet diameter at the base. In commemoration of the passing of the Reform Act. 2 sheets Lithograph. London, 1832.

1833 BRUNEL, SIR MARC ISAMBARD. *The Thames Tunnel: an exposition of facts submitted to the King by Mr Brunel.* London, 1833, 4to. Gives an account of the driftway based obviously on first-hand information.

1836 HEBERT, LUKE. *The Engineer's and Mechanic's Encyclopaedia.* London, 1836, 8vo. Trevithick's engine and locomotive described, pp. 387, 734.

1839 CLARKE, HENRY HYDE. "Memoir of Richard Trevithick." *Civil Engineers' and Architects' Journal,* Vol. II, p. 93. London, 1839, 4to. The author knew Trevithick during a most active portion of his career and was his first biographer.

c. 1840 FAREY, JOHN. *Treatise on the Steam Engine,* Vol. II. London, c. 1840. Although put into type, Vol. II of this work was not published, and is known only through a proof in the possession of the Patent Office Library. Trevithick's work on the high-pressure engine is given, pp. 7–42.

1844 POLE, WILLIAM, F.R.S. *A Treatise on the Cornish Pumping Engine.* London, 1844, 4to. Valuable because of the writer's standing and because he recorded information obtained at first-hand in Cornwall. Refers to Trevithick, pp. 32, 45, 162, and Appendix G.

1847 CLARKE, HENRY HYDE. "The high pressure steam engine and Trevithick." *Railway Register*, Vol. v. London, 1847, 8vo. A review, pp. 73–96, of Dr Alban's *High Pressure Steam Engine*, trans. by William Pole, 1847, in the course of which Clarke, pp. 86–96, gives "A short sketch" of Trevithick's career, subsequently reprinted without date under the title *Life of Richard Trevithick, C.E.* 12mo. It was reprinted in *Mech. Mag.* Vol. XLVI, 1847, p. 301.

1849 CLARKE, HENRY HYDE. "Life of Richard Trevithick, C.E." *Mining Almanack*, ed. by Henry English, p. 303. London, 1849, 8vo. Practically the same matter as the preceding.

1857 LAW, HENRY. *Memoir on the several operations and the construction of the Thames Tunnel.* London [1857], 4to. Gives the best account we have of the driftway, Pt I, pp. 3–7.

1859 EDMONDS, RICHARD, JUNIOR. "Contribution to the biography of Richard Trevithick." *Edinburgh New Philosophical Journal*, Vol. x, new series, p. 327. Edinburgh, 1859, 8vo. Edmonds was the son of Richard Edmonds, Trevithick's solicitor and friend; the facts adduced were within his personal knowledge.

1862 EDMONDS, RICHARD. *The Land's End District...also a brief memoir of Richard Trevithick.* London, 1862, 8vo. With a map. Comprises much the same matter as the preceding, expanded.

1862 WALKER, WILLIAM. *Memoirs of distinguished Men of Science.* London, 1862, 8vo. Short Memoir, pp. 193–200. Published in connection with his well-known picture.

1862 BEAMISH, RICHARD, F.R.S. *Memoir of the life of Sir Marc Isambard Brunel.* London, 1862, 8vo. Gives a fair account of the Thames driftway, pp. 203–206.

1872 TREVITHICK, FRANCIS, C.E. *Life of Richard Trevithick, with an account of his inventions.* London, 1872, 2 vols. 8vo. The writer was the son of the great engineer. The source book of nearly all we know about him and very extensively used in this work, where it is quoted briefly as *Life.* The writer, unwisely, edited severely the letters he quoted. Wherever possible the authors have given the letters as written, except for necessary punctuation.

1878 BOASE, G. C. and COURTNEY, W. P. *Bibliotheca Cornubiensis.* London, 1878, 4to. Trevithick's biography, bibliography, and patent specifications.

1883 DAVIS, MAJ.-GEN. JOHN. *Memorial edition of the life of Richard Trevithick.* London, 1883, 8vo. An epitome of the *Life* above, prepared in aid of the appeal for funds for the Jubilee Memorial of Trevithick.

1884 TREGELLAS, WALTER. "Trevithick the Engineer" in *Cornish Worthies*, Vol. II, pp. 305–344. London, 1884, 8vo. Based on the *Life* above, but treated mainly from the human side.

1890   BOASE, G. C. *Collectanea Cornubiensis.* London, 1890, 4to. Genealogy of the Trevithick family, p. 1090.

1899   SECCOMBE, THOMAS. *Dictionary of National Biography.* Art. "Trevithick, Richard". London, 1899, 8vo. Best condensed account of his life.

1902   BERINGER, JOHN JACOB, A.R.S.M., F.I.C. *Richard Trevithick*, pp. 15–35.

1902   KEAST, J. CHAMPION. *Mechanical Engineering in the days of Trevithick.* pp. 36–49.

1902   THOMAS, WILLIAM, M.Inst.C.E. *Cornish Mining in the time of Trevithick*, pp. 50–65. Camborne, 1902, 8vo. These three papers were published in connection with the Centenary demonstration of the Camborne steam carriage on Christmas Eve, 1901. Edited by J. J. Beringer, principal of the Camborne School of Mines.

1909–1910   Fox, HOWARD, F.G.S. "Boulton and Watt" and "Boulton and Watt (No. 2)." *Reports of Royal Cornwall Polytechnic Society*, 1909 *and* 1910. Penryn, 1910 and 1911, 8vo. Extracts from and comments on the *Boulton and Watt Letters* (see above).

1913   HARPER, EDITH K. *A Cornish Giant; Richard Trevithick, the father of the locomotive.* London, 1913, 8vo. An appreciative sketch; contains hitherto unpublished matter.

1927   JENKIN, A. K. HAMILTON, M.A., B.Litt. "Boulton and Watt in Cornwall." *Reports of Royal Cornwall Polytechnic Society*, 1926. Camborne, 1927, 8vo. Extracts from and comments on the *Boulton and Watt Letters* (see above).

1927   DICKINSON, H. W. and JENKINS, RHYS. *James Watt and the Steam Engine.* Oxford, 1927, 4to. Relations of Trevithick in his early manhood with Boulton and Watt, p. 310.

1928   TITLEY, ARTHUR. "Trevithick and Rastrick and the single-acting expansive engine." *Trans. Newcomen Soc.* Vol. VII, pp. 42–59. Leamington, 1928, 4to. Exhaustive study of this invention.

1929   CARAH, REV. CANON J. SIMS. "Richard Trevithick and his Home at Penponds." *Cornish Almanack*, pp. 29–37 Excerpt. Camborne, 1929, 12mo. Personalia and particulars of the house.

1931   TITLEY, ARTHUR. "Richard Trevithick and the Winding Engine." *Trans. Newcomen Soc.* Vol. X, pp. 55–68. Leamington, 1931, 4to. Survey of his contribution to winding in mines.

1933   INGLIS, CHARLES EDWARD, O.B.E., M.A., F.R.S., M.Inst.C.E. Trevithick Memorial Lecture. London, 1933, fol.

The Pedigree of the Trevithick family between pages 284 and 285 has been added mainly in order to show what a large number of Trevithick's descendants have followed engineering or an allied profession.

This Pedigree is now available for download from www.cambridge.org/9781108016353

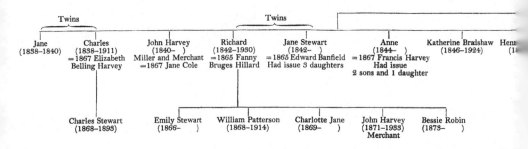

Richard (1798–   )
d. unmarrio[...]

Twins

| | Twins |

Jane (1838–1840)

Charles (1838–1911)
=1867 Elizabeth Belling Harvey

John Harvey (1840–   )
Miller and Merchant
=1867 Jane Cole

Richard (1842–1930)
=1865 Fanny Bruges Hillard

Jane Stewart (1842–   )
=1865 Edward Banfield
Had issue 3 daughters

Anne (1844–   )
=1867 Francis Harvey
Had issue
2 sons and 1 daughter

Katherine Bradshaw (1846–1924)

Henr[...] (1[...]

Charles Stewart (1868–1893)

Emily Stewart (1866–   )

William Patterson (1868–1914)

Charlotte Jane (1869–   )

John Harvey (1871–1933)
Merchant

Bessie Robin (1873–   )

PEDIGREE OF TREVITHICK FAMILY

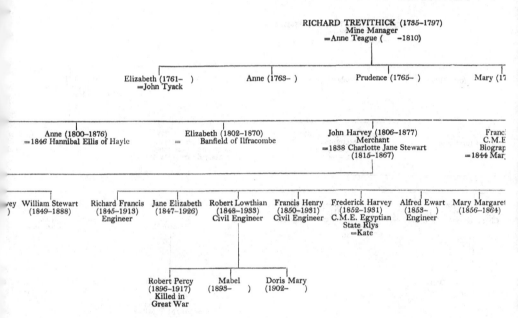

RICHARD TREVITHICK (1735–1797)
Mine Manager
=Anne Teague (–1810)

Elizabeth (1761–   )   Anne (1763–)   Prudence (1765–)   Mary (17
=John Tyack

Anne (1800–1876)   Elizabeth (1802–1870)   John Harvey (1806–1877)   Franc
=1846 Hannibal Ellis of Hayle   =   Banfield of Ilfracombe   Merchant   C.M.E
                                                              =1838 Charlotte Jane Stewart   Biograp
                                                                        (1815–1867)   =1844 Mar

vey William Stewart   Richard Francis   Jane Elizabeth   Robert Lowthian   Francis Henry   Frederick Harvey   Alfred Ewart   Mary Margaret
)   (1849–1888)   (1845–1913)   (1847–1926)   (1848–1933)   (1850–1931)   (1852–1931)   (1853–)   (1856–1864)
                                 Engineer   Civil Engineer   Civil Engineer   C.M.E. Egyptian   Engineer
                                                                                                      State Rlys
                                                                                                      =Kate

                       Robert Percy   Mabel   Doris Mary
                       (1896–1917)   (1893–   )   (1902–   )
                       Killed in
                       Great War

PEDIGREE OF TREVITHICK FAMILY

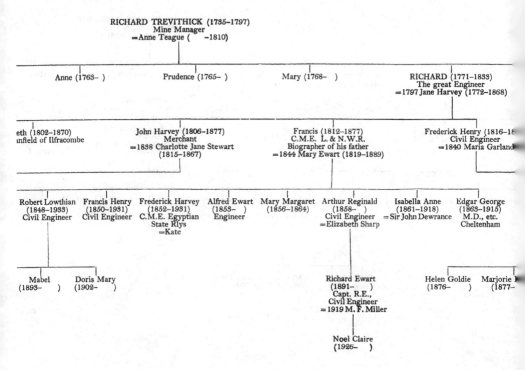

RICHARD TREVITHICK (1735–1797)
Mine Manager
=Anne Teague (–1810)

Anne (1763–)   Prudence (1765–)   Mary (1768–)   RICHARD (1771–1833)
The great Engineer
=1797 Jane Harvey (1772–1868)

...eth (1802–1870)
...nfield of Ilfracombe

John Harvey (1806–1877)
Merchant
=1838 Charlotte Jane Stewart
(1815–1867)

Francis (1812–1877)
C.M.E. L. & N.W.R.
Biographer of his father
=1844 Mary Ewart (1819–1889)

Frederick Henry (1816–1...
Civil Engineer
=1840 Maria Garland...

Robert Lowthian
(1848–1933)
Civil Engineer

Francis Henry
(1850–1931)
Civil Engineer

Frederick Harvey
(1852–1931)
C.M.E. Egyptian
State Rlys
=Kate

Alfred Ewart
(1853–)
Engineer

Mary Margaret
(1856–1864)

Arthur Reginald
(1858–)
Civil Engineer
=Elizabeth Sharp

Isabella Anne
(1861–1918)
=Sir John Dewrance

Edgar George
(1863–1915)
M.D., etc.
Cheltenham

Mabel
(1893–)

Doris Mary
(1902–)

Richard Ewart
(1891–)
Capt. R.E.,
Civil Engineer
=1919 M. F. Miller

Helen Goldie
(1876–)

Marjorie...
(1877–

Noel Claire
(1926–)

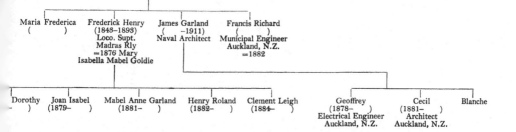

Thomasina
= Henry Vivian

Maria Frederica   Frederick Henry   James Garland   Francis Richard
( )   (1843–1893)   (–1911)   ( )
    Loco. Supt.   Naval Architect   Municipal Engineer
    Madras Rly   Auckland, N.Z.
    =1876 Mary   =1882
    Isabella Mabel Goldie

Dorothy   Joan Isabel   Mabel Anne Garland   Henry Roland   Clement Leigh   Geoffrey   Cecil   Blanche
)   (1879–)   (1881–)   (1882–)   (1884–)   (1878–)   (1881–)
            Electrical Engineer   Architect
            Auckland, N.Z.   Auckland, N.Z.

INDEX

The figures in heavy type refer to illustrations

Abadia, Don Pedro, 159, 160, 168, 176, 186
Adventurers, mine, 8
Agricultural engines, 130–4
Archway, Thames, 90–105, 186
Arismendi, Don José, 159, 160, 168, 176, 187

Ballast heaving, 84, 89
Barrow, Sir J., Secretary to the Admiralty, 242
Beacon Hill, Trevithick statue erected near, 47
Bendy, T., 82
Bentham, Gen. Sir S., 80
Bill, W., 113
Birkbeck, Dr G., 208
Births register, Illogan, **13**
Blackett, C., 70
Blast pipe locomotive, 66
"Blazer", gun brig, 81, 89
Blewett, G., 121, 135
Boiler, Cornish, **4**, 125, 127, 207, 209; Lancashire, **4**; cast-iron, 60, 61; direct-heated, **126**; water-tube, **152**; tubular, 242, **244**
Bolivar, S., 185, 187, 197
Boulogne, expedition against, 77
Boulton, M., **14**, 15, 23, 28
Boulton, M. R., 24
Boulton & Watt, come to Cornwall, 14; contests with Trevithick, 22–35; espionage by and upon, 27; *v.* Hornblower & Maberley, 28, 35; leave Cornwall, 35
Bramah, J., 219
Branca, steam turbine of, 2
Brunel invents tunnelling shield, 105
Bull, E., 20; engines of, **21**; proceeded against by Boulton & Watt, 22, 28, 31, 32
Bull, W., 169, 185
Burials Register, Dartford, **256**

Camborne road carriage, 44, **49**, 50–2
Catch me who can, **108**

Cerro de Pasco Mines, 159, 160, **172**, 174, 181, 185, 187
Chapman, W., engineer, 92
Clarke, Hyde, biographer, 264, 282, 283
Closed cycle steam engine, 236–42
Coals, charges upon, 251
Cochrane, Sir T., 185, 216, 217
Company promotion, 215
Compressed air transmission, 217
Cook's Kitchen Mine, whim engine at, **45**
Copper mining, Chile, 188
Cornish boiler, *see* Boiler
Cornish engine, first, 129
Cornish engines, performances of, 229–34
Cornwall, characteristics of, 6; tin mining in, 7 ff.
Cost Book system of mining, 7–9
Creighton, W., assistant to Watt, 235
Cugnot, N. J., builds steam carriage, 46
Cultivator, steam, 133
Cunningham, Lt., 68
Curtis, Sir W., 84

Dartford, experiments at, 254; death of Trevithick at, 256
Davy, Sir Humphry, 50, 118
Day, *see* Page
Dedunstanville, Lord, purchases thrashing engine, 131
Dennis, Mrs, reminiscences of, 23
Deverill, Mr, 81, 83
Dickinson, R., inventor, exploits Trevithick, 114–20
Docks in London, excavating at, 83
"Dragons", Trevithick's, 51
Dredgers, 80, **86**
Dredging, steam, 80–90
Dundonald, Earl of, 185, 216, 217
Duty of steam engines, 18, 229–31, 233

Edmonds, R. (jun.), 16, 37
Edmonds, R. (sen.), solicitor to Trevithick, 175, 209, 210, 211, 225

Edmonds, T., 189, 200
Edwards, Mr, 167
Explosion at Greenwich, 59–61

Falmouth Harbour, 6; packets, 7, 122
Farey, J., 60, 61, 208, 235
Felton, W., 56, 57
Field, J., 135
Fox, Messrs, 31, 32
Fox, Williams & Co., contractors, 138

Gerard, J. M., adventures with Trevithick, 196–206; partner with Trevithick in proposed Company, 209–13; death, 214
Giddy, Davies, changes name to Gilbert, 36, 109; consulted by, and correspondence with, Trevithick, *passim*
Gittins, W., school of, 213
Goodrich, S., 56, 65, 68, 104, 195, 242, 246
Green, J., 135, 136
Greenwich, explosion at, 59–61
Grose, Capt. S., 143, 158, 230
Guilmard, Mrs, sister to Giddy, 107, 108
Gundry, Capt. T., 25
Gun-mounting, recoil, 215, **216**, 217
Gwennap, Wesley memorial service held annually at, 12

Hall, B. M., rescues Trevithick, 204
Hall, E., patents recoil engine in France, 255
Hall, J., engineer at Dartford, 254, 255
Hambly, S., carpenter, 118
Harvey, H., brother-in-law to Trevithick and proprietor of Hayle Foundry, 121, 145, 211, 217
Harvey, Jane, 37, 44, 190–5, 210, *see* Trevithick
Harvey, N., 220
Hawkins, J. I., 4 *n.*
Hawkins, Sir C., Bt., 126, 131, 134
Hayle Foundry, 55, 135, 136, 217
Haynes & Douglas, cotton merchants, 109, 113
Hazledine, Rastrick & Co., 76, 135, 164, 165, 168
Heating, hot water, **248**, 249, 250

Herland engine, 144–7, 226, 235
Heron of Alexandria, aeolipile of, 2, 149
High-pressure engine and boiler, 1, 43, **62**
High-pressure patent, 53; disposal of, 113; specification, 271
Hodge, Capt., 18, 34, 35, 184, 187
Holland, drainage scheme for, 220–3, **224**, 225
Holman, J. M., 266
Homfray, S., 61, 63, 68, 72, 75, 109, 110, 113, 208
Hornblower, Jonathan, 11, 18
Hornblower, Joseph, 11
Hosking, R., describes Wheal Prosper engine, 127–9
Hughes, Bough & Mills, 81, 82
Hutton, Dr, mathematician, 105
Hydraulic power transmission, 219

Ice making, 219
Illogan parish, Trevithick born in, 13
Invasion of England, 77
Invention, 1–3
Iron tanks, 118–19, 122

Jessop, W., civil engineer, 83, 105

Kendal, Mr, 131
Kevil, Mr, Watt rejects proposal of, 25

Lean, Capt. J., "Registrar and Reporter" on engine performances, 229
Letters Patent, **53**, 269–80
Liddell, Lt. J., meets Trevithick at Callao, 189
Lima, *Government Gazette*, extract from, 172
Limestone quarrying, 139
Linnell, J., *see* Trevithick portraiture
Linthorn, Mr, 217, 219
Liverpool and Manchester Railway commenced, 207
London locomotive, 105–13
London road carriage, 55–9

Maberley lawsuit, 28, 35
Maudslay, H., acquires patent for iron tank, 119, 246

Melville, Lord, 78, 79
Merrifield, L. S., *see* Trevithick portraiture
Michell, W. A., 13
Miers, J., 175, 176, 177, 184, 186
Mills, G., writes to Admiralty and Royal Society on Trevithick's behalf, 246
Mills, J., shareholder in Peruvian mines, 156, 168
Mine administration, 8, 9
Mine(s):
  Alport, **41**, 42
  Balcoath, 21
  Bedworth, 20
  Binner Downs, 38, 231, 234
  Camborne Vean, 38
  Caxatamba, 182, 183, 194, 206, 215
  Conchucos, 184
  Consolidated, 227, 230
  Cook's Kitchen, 19, 38, **45**
  Crane, 19
  Ding Dong, 22, 23, 30, 32, 34, 38
  Dolcoath, 12, 38, 42, 85, 126, 127
  East Pell, 38
  Eastern Stray Park, 12, 16
  Godolphin, 38
  Hallamannin, 22, 38, 46
  Herland, 144, 145, 226, 227, 230, 235
  North Downs, 52
  Padre Arias, 198, 199
  Pasco, 159, 160, 172, 180–2, 184–7
  Pednandrea, 32
  Penberthy Crofts, 75, 76
  Penrose, 38
  Poldice, 22
  Poldory, 148
  Polgine, 38
  Prince William Henry, 12, 31, 38, 39
  Quebrada-honda, 198, 199, 206
  Relistan, 129
  Rosewall, 38
  Roskear, 12, 31, 39
  St Agnes, 25, 33, 34
  Seal Hole, 18, 38
  Tincroft, 18
  Tregonick, 157
  Trenethick St Agnes, 38
  Trenethick Wood, 38, 41

Mine(s): (*cont.*)
  Treskerby, 148, 210, 211
  Trevenen, 38
  United, 28, 29, 145, 209–11
  Wheal Abraham, 34, 38
  Bog, 38
  Chance, 12, 31, 148, 210, 211
  Clowance, 38
  Crenver, 30
  Damsel, 148, 231
  Druid, 39, **40**, 41
  Francis, 158, 191
  Gons, 20, 38
  Hope, 38
  Leeds, 34, 35
  Maid, 45
  Malkin, 23, 38
  Margaret, 38
  Prosper, 38, 127, **129**, 130, 143
  Providence, 23
  Rose, 38
  Towan, 230–2
  Treasury, 12, 20, 27, 30, 38, 46
  Union, 15
  Unity, 30
  Vor, 10, 145, 231
Mining in Cornwall, 7
Mint at Lima, 170, 177, 182
Mitchel, R., 28
Montelegre boys, 202, 205, 213, 214
Morcom, Capt. J., 18
Morland, Sir S., first patentee of plunger pump, 38
Murdock, W., 20, 23, 25–7, 30, 31, 46, 219

Nautical labourer, 115, 116, **117**
Navigation, steam, 79, 115, **116**, **142**, 149, 151, 156, 157, 241, 247
Neath Abbey Ironworks, 162, 164
New Improved Patent Steam Navigation Co., 247
Newcastle locomotive, **70**
Newcomen, T., 2, 10
Nicaragua, journey across, 200, **201**, 202, 203

Oats, J., boilermaker, 143, 156

Paddle steamer, 79, **116**, 149
Page, R., lawyer, 168, 175–81
Parliament, petition to, 225–9

Patent, Letters, **53**, 269–80
Penn, J., father of the marine engineer, 157
Penponds, Trevithick's home at, 14, **122**
Penydaren locomotive, 62–70
Perkins, J., 207
Peru, silver mines in, 159
Phillips, H., reminiscences of, 145, 147
Pickwood, R. W., 135, 170
Plug and Feathers for breaking up stone, **138**
Plunger-pole engine, 142–5, **146**, 147, 148, 153, **154**
Plymouth Breakwater, 136–40
Pole, Dr W., 129
Preen, R., reminiscences of, 153
Price, J., 145, 163
Pump, plunger, 38; plunger-pole, 142–5, **146**, 147, 148, 209–11; ball and chain, 223, **224**, 225

Quicksand, tunnelling through, 99

Rastrick, J. U., colleague of Trevithick, 79, 110, 131, 135; tour in Cornwall, 136–42; 145, 156, 161, 162, 163, 166; railway engineer, 208
Recoil engine, 149, **150**, 151, 153, 155
Reform Act column, 252, **253**, 254
Rennie, G., 240, 241
Rennie, J., engineer, 92; recommends breakwater and submits plans for Plymouth, 136
Robinson & Buchanan, 132
Rock boring, 138, 139
Rolling mill, **71**, 169
Rowe, S., 118
Rowlandson drawing of London railway, 111, 113
Rowley, W., engine maker, 82, 160
Rumford, Count, 50

San Francisco Harbour, tunnel across, 105
Sanders, or Saunders, J., boilermaker, 175, 191
Savery, T., engine of, 2
Science Museum, South Kensington, 44, 131, 261

Screw propeller, 155–7
Sims, W., engineer of the Eastern Mines, 148, 192, 209
Sinclair, Sir J., 69, 106; first President of the Board of Agriculture, 134, 149
Smith, J., millwright, 156, 157, 168, 177, 192
Southern, J., assistant to Watt, 235
Sporting event, 107
Spry, R., 185, 186
Stafford, Marquis of, 78, 79
Stannary jurisdiction, 7, 9
Steam, properties of, 235
Steamboats, 79, 115, **116**, **142**, 149, 151, 156, 157, 241, 247
Steam carriage, Camborne, 44, 46–52; London, 55–9; Select Committee on, 251
Steam cultivator, **133**
Steam engine, invention of, 3; high-pressure, 4, 42, 43–5, **62**, **117**, 125, 135, 160, **271**, **274**; atmospheric, 10; Bull's, 20, 21, 25; "Topsey-Turvey," 21; dredger, 80–90; Cornish, first, 129; thrashing, 130–4; sugar mill, 134, **274**; plunger-pole, 142, **146**, 154; compound, 148, 149; recoil, 155, 238; winding, for S. America, **166**; duty of, 229–34; closed-cycle, **237**. *See also* Trevithick
Steam road carriage, **277**
Steamship propulsion, 79, 115, **116**, **142**, 149, 151, 155–7
Steel, J., 70, 118, 121
Stephenson, R., 205
Stevenson, W. B., 184
Stove, warming, 248–50
Sugar mills, 134, 274
Surgery, amateur, 189

Tanks, iron, 118–19, 122
Teague, Anne, 12; death, 122
Teague, J., 161, 163, 167, 168
Teague, W., 192
Thames Archway Co., 90–105, 186; *v.* Chapman, 109
Thames Driftway, 92–104, **105**
Thrashing engine, **130**
Ticketing of ores, 10, 12
Tincroft engine trial, **18**
Tramway locomotive, 62–8, **69**, 70

Tredegar rolling mill engine, **71**, 72, **73**

Tregajorran, reputed birthplace of Trevithick, 14

Trevarthen, T., 169

Trevithick Memorial Committee, 262

Trevithick, A. (*née* Teague), 12

Trevithick, F., 12, 13, 98, 111, 120, 123, 175, 183, 192, 208, 222, 225, 230, 261, 264

Trevithick, H. H., 261

Trevithick, R. (sen.), 12; death, 31

Trevithick, R., guiding principle of life, 5; birth, 13; birthplace, 13; schooling, 14, 16; boyhood, 16; chooses career, 16; first employment, 16; feats of strength, 19; meets Edward Bull, 20, 21; Mrs Dennis's reminiscences of, 23; contests with Boulton & Watt, 22–35; not partner with Bull, 31; meets Giddy, 36; marriage, *see* Harvey, Jane; employed as engineer, 37, 38; experiments with high-pressure steam, 38–59; describes boiler explosion at Greenwich, 59, 60; locomotive experiments, 62–70; demand for his stationary engine, 70–6; volunteers for home defence, 76–7; impresses Marquis of Stafford, 78; steamboat trials, 79; experiments with dredgers, 80–9; interested in Thames Archway, 90–105; joined by family in London, 98; his London locomotive, 106–14; exploited by Dickinson, 114–20; incapacitated by illness, 120–22; bankrupt, 122–4; returns to Cornwall, 122; starts life afresh, 125; expansive working with steam, 126–33; supplies engines for West Indies, 134, 135; work on Plymouth Breakwater, 136–40; erects plunger-pole engine, 142; rivalry with Woolf, 143–8; experiments with recoil engine, 149–57; dealings with Uvillé, 161–75; sails for Peru, 175; jealousy and intrigue, 180–1; prospecting in Chile, 182–4; destruction of mines by revolutionaries, 185, 186; wife's devotion to, 190–5; leaves Costa Rica for Caribbean Sea with Gerard, 200–3; meets Robert Stephenson, 203; returns to England, 206; changes during absence of, 207; proposes formation of Company with Gerard, 209–15; designs gun carriage, 215–17; visits Holland, 220; petitions Parliament for grant, 225; observations on performances of engines, 231–40; applies to Admiralty for vessel, 241; takes out patents for boilers and superheaters, 243, 247; invents hot-water stove, 248–50; proposes and designs Reform Act column, 252–4; illness, death and burial, 255; *see also* Dartford; character of, 257–60; portraiture of, 261–3; memorials to, 264–7; letters patent, **53**, 269–80; correspondence with Giddy, *passim*

Trevithick, R. (son), 118

Trinity House, Corporation of, 84–9

Tunnel under Thames, 90–104, **105**, 186

Tyack, Elizabeth, sister of Trevithick, 190

Uvillé, Francisco, 159–65, 168, 169–71, 173, 175, 177, 181; death, 185

Vazie, R., engineer, 90, 95, 96

Vivian, Capt. A., 24, 50, 51, 52, 55, 56, 62, 145, 158

Vivian, H., 48, 57, 169; death, 177

Vivian, Capt. J., 57, 83

Vivian, Capt. N., 127

Vivian, S., 23, 24

Water-jet propulsion, 247

Water-pressure engine, 39–42

Watkin, Sir E., 204, 205

Watt, J. (jun.), 4 *n*., 5, 23, 24, 25, 29, 235

Watt, J. (sen.), separate condenser, 3, 14; importance of his engine to Cornwall, 15; patent challenged, 17; opposes Bull, 21; secures injunction against Trevithick, 23. *See also* Boulton & Watt

Weighing machine for coals, 251

Wesley, J., on Cornwall, 12

West, W., 44, 133

Wheal = Mine, *q.v.*
Wheal Maid, Watt erects engine at,
  45
Whinfield, J., 70
Whitehead & Co., 44
Williams, J., 210, 211, 212
Williams, W., 188
Wilson, T., 20, 24
Winding engines, **45, 166**

Woolf, A., rival to Trevithick, 143–8,
  149, 195, 230
Worcester, experiment of Marquis
  of, 2
Wreck raising, 119
Wrestling, Cornish, 19
Wylam Colliery, 70

Zuider Zee, drainage of, 222

CAMBRIDGE: PRINTED BY W. LEWIS, M.A., AT THE UNIVERSITY PRESS

Printed in the United States
By Bookmasters